KU-180-201

COMBINATORIAL OPTIMIZATION

Annotated Bibliographies

COMBINATORIAL OPTIMIZATION

Annotated Bibliographies

edited by

M. O'hEigeartaigh
National Institute for Higher Education, Dublin

J.K. Lenstra
Centre for Mathematics and Computer Science, Amsterdam

A.H.G. Rinnooy Kan
Econometric Institute, Erasmus University, Rotterdam

A Wiley-Interscience Publication

CENTRE FOR MATHEMATICS AND COMPUTER SCIENCE
Amsterdam

JOHN WILEY & SONS
Chichester · New York · Brisbane · Toronto · Singapore

The Centre for Mathematics and Computer Science is a research institute of the Stichting Mathematisch Centrum, which was founded on February 11, 1946, as a nonprofit institution aiming at the promotion of mathematics, computer science, and their applications. It is sponsored by the Dutch Government through the Netherlands Organization for the Advancement of Pure Research (Z.W.O.).

Copyright © 1985 by John Wiley & Sons, Ltd.

All rights reserved.

No part of this book may be reproduced by any means, nor transmitted, nor translated into a machine language without the written permission of the publisher.

Library of Congress Cataloging in Publication Data:

Main entry under title:

Combinatorial optimization.

'A Wiley-Interscience publication.'
1. Combinatorial optimization — Bibliography.
I. O'hEigeartaigh, M. II. Lenstra, J.K.
III. Rinnooy Kan, A.H.G. IV. Title.
Z6654.M28C66 1985 [QA402.5] 519 84-5081
ISBN 0 471 90490 2

British Library Cataloguing in Publication Data:

Combinatorial optimization.
1. Combinatorial optimization — Bibliography
I. O'hEigeartaigh, M. II. Lenstra, J.K.
III. Rinnooy Kan, A.H.G.
016.511'6 Z6654.C6/

ISBN 0 471 90490 2

Photosetting by the Centre for Mathematics and Computer Science, Amsterdam
Printed in Great Britain

Table of Contents

Preface

This book is the result of a summer school on combinatorial optimization that was held in Dublin from July 4 until July 15, 1983. During these twelve days, ten invited lecturers presented surveys as well as summaries of their own recent work. The program also included several short presentations by the participants.

The summer school, which had an international attendance of 76 people, demonstrated the rapid rate at which the area of combinatorial optimization is developing. The lectures touched upon discrete mathematics and graph theory, but also on probability theory and statistics, and were influenced by developments in various branches of computer science.

This expansion, involving so many disciplines, has made it difficult to maintain an overview of the literature. The purpose of this book is to facilitate this process for all those with an interest in combinatorial optimization. The format chosen is such that the book occupies the middle ground between a pure bibliography and a collection of surveys. For twelve subareas of combinatorial optimization, specialists have been asked to select the main references and to provide brief annotations. Important recent publications that at the time of writing had only appeared as technical reports are also included. The dates at which the contributions were completed range from June 1983 to July 1984.

Of the twelve bibliographies, nine were prepared by lecturers at the summer school and three (Chapters 7, 9, and 12) were solicited thereafter. We are grateful to the contributing authors for the quick way in which they have produced their manuscripts.

A special word of thanks is due to the National Institute for Higher Education in Dublin and, in particular, to its Director, Dr. D. O'Hare, for supporting the conference that laid the foundation for this book. We would also like to thank the other members of the organizing committee, R. Faulkner, C.S. Edwards, J. Horgan, M. Jordan, A. Moynihan, C.N. Potts, M. Ryan, M. Scott, and I. Williams, for their efforts in making the conference a success.

All material has been typeset on the UNIX® system at the Centre for Mathematics and Computer Science (CWI) in Amsterdam. We owe many thanks to the staff of the CWI, especially to Sandra Dorrestijn and Aafje van den Berg, for their helpful efficiency.

M. O'hEigeartaigh
J.K. Lenstra
A.H.G. Rinnooy Kan

1

Polyhedral Combinatorics

M. Grötschel
Universität Augsburg

The subject *polyhedral combinatorics* has grown so explosively in the recent twenty years that it is virtually impossible to write an (annotated) bibliography of this field aiming at a high degree of completeness. This is rather easy for the period before 1970 when the field had not yet completely emerged as a subject in its own right. There are classics like

L.R. Ford, D.R. Fulkerson (1962). *Flows in Networks,* Princeton University Press, Princeton, NJ,

which is a landmark book in network flow theory, like

G.B. Dantzig, D.R. Fulkerson, S.M. Johnson (1954). Solution of a large-scale traveling-salesman problem. *Oper. Res. 2,* 393-410,

which describes the basic techniques of cutting plane generation and separation within a linear programming framework (although not totally explicit and aware of all relations to polyhedral theory, but many of the fundamental ideas can be found there), like

A.J. Hoffman, J.B. Kruskal (1956). Integral boundary points of convex polyhedra. [Kuhn & Tucker 1956] (see below), 223-246,

which introduces totally unimodular matrices (called matrices with the unimodular property there) and recognizes them as an important class of matrices in polyhedral combinatorics with many applications, or like

J. Edmonds (1965). Maximum matching and a polyhedron with 0,1-vertices. *J. Res. Nat. Bur. Standards Sect. B 69,* 125-130,

J. Edmonds (1965). Paths, trees, and flowers. *Canad. J. Math. 17,* 449-467,

where the first complete linear description of a polytope associated with a combinatorial optimization problem (matching) is given which is nontrivial (in the sense that it needs more inequalities, in fact an exponential number, for the description than are necessary for the obvious integer programming formulation) and where a polynomial time algorithm based on linear programming techniques is presented utilizing this description, showing for the first time that structure (nice characterization of the inequalities) matters more than number (of inequalities) for the design of good algorithms. In particular this and the subsequent work of Edmonds (and coauthors), for instance

J. Edmonds (1970). Submodular functions, matroids, and certain polyhedra. R. Guy *et al.* (eds.). *Combinatorial Structures and their Applications,* Gordon and Breach, New York, 69-87,

J. Edmonds, E.L. Johnson (1970). Matching: a well-solved class of integer linear programs. *Ibidem,* 89-92,

J. Edmonds, R. Giles (1977). A min-max relation for submodular functions on graphs. *Ann. Discrete Math. 1,* 185-204,

has influenced the development of polyhedral combinatorics to a great extent.

It seems that even the work on special subjects like polyhedral aspects of the traveling salesman problem or stable sets in graphs or on submodular flows has grown so rapidly that it may take almost a book to survey these topics in detail. Fortunately, a number of very good and up to date survey papers on various aspects of polyhedral combinatorics have been written recently which treat the existing literature in depth. For those who wish to get acquainted with polyhedral combinatorics we recommend reading (some of) those papers we will describe in the sequel. Most of these papers have a large number of references which make it easy to get to the present research literature and approach the frontiers of this science.

Before starting our *survey of surveys* we would like to mention a project carried out at the Institut für Operations Research, Universität Bonn, Nassestrasse 2, D-5300 Bonn, West Germany. The Institute collects and compiles all papers and books on *Integer Programming and Related Areas.* Classified bibliographies are published in the Lecture Notes in Economics and Mathematical Systems of Springer Verlag about every three or four years containing almost everything published in this area during the covered time period. The first three books of this series are the following:

C. Kastning (ed.) (1976). *Integer Programming and Related Areas: a Classified Bibliography,* Lecture Notes in Economics and Mathematical Systems 128, Springer, Berlin,

D. Hausmann (ed.) (1978). *Idem 1976-1978,* Ibidem 160, Springer, Berlin,

R. von Randow (ed.) (1982). *Idem 1978-1981,* Ibidem 197, Springer, Berlin.

The papers and books are classified according to a subject list. Papers and books on aspects of polyhedral combinatorics can be found for instance under the following subjects:

- adjacency on integer polyhedra;
- blocking, antiblocking, integer rounding;
- cutting planes;
- duality in integer programming;
- facets of integer polyhedra;
- group theoretic approach (corner polyhedra);
- integer polyhedra.

These bibliographies are valuable sources of literature on combinatorial optimization and make the search for existing papers on a subject quite easy.

We now give an overview on papers which survey various aspects of polyhedral combinatorics. Many of these papers have been prepared for tutorial purposes and outline the subject they treat in a way that is also suitable for the beginner.

Polyhedral combinatorics deals with the application of polyhedral theory and linear algebra to combinatorial problems. There are only very few books on polyhedral theory (compared to the vast amount of linear algebra books), and moreover, most of these books do not treat all those polyhedral topics which are of interest in polyhedral combinatorics. The same is also true for most of the books on linear programming. Two books - although dealing with more general subjects - which are good sources for the type of polyhedral theory we need are:

J. Stoer, C. Witzgall (1970). *Convexity and Optimization in Finite Dimensions I,* Grundlehren der mathematischen Wissenschaften 163, Springer, Berlin,

R.T. Rockafellar (1970). *Convex Analysis,* Princeton University Press, Princeton, NJ.

A comprehensive survey of the part of polyhedral theory used in mathematical programming is

A. Bachem, M. Grötschel (1982). New aspects of polyhedral theory. B. Korte (ed.). *Modern Applied Mathematics: Optimization and Operations Research,* North-Holland, Amsterdam, 51-106.

A classic collection of papers on various aspects of polyhedral theory and combinatorics is

H.W. Kuhn, A.W. Tucker (eds.) (1956). *Linear Inequalities and Related Systems,* Princeton University Press, Princeton, NJ,

where many of the results of the algebraic theory of polyhedra (which today are mostly considered as basic knowledge) appear for the first time. Most of the papers in this book are still worth reading, not only for historical interest.

Many of the survey papers mentioned in the sequel also give a short overview over the particular part of polyhedral theory necessary for the paper. In the near future the following book will appear that contains all those areas of linear algebra and polyhedral theory (theoretical as well as algorithmical aspects) which are of interest in combinatorial optimization:

A. Schrijver (to appear). *Polyhedral Combinatorics,* Wiley, Chichester.

At present, the results of polyhedral combinatorics are scattered over the literature, and only a few books have a chapter on special topics of it. The forthcoming book by Schrijver will be an exceptionally comprehensive treatment of polyhedral combinatorics and cover almost all parts of the theory known to date. We highly recommend this book (parts of which are circulating in the scientific community) to students and researchers. The broadest and most up to date survey paper covering the whole field is

W.R. Pulleyblank (1983). Polyhedral combinatorics. A. Bachem, M. Grötschel, B. Korte (eds.). *Mathematical Programming: the State of the Art - Bonn 1982*, Springer, Berlin, 312-345.

The paper starts with basic polyhedral theory, deals with faces, facets, dimension and adjacency, various representations, dual integrality, discusses prominent examples (e.g. matching polyhedra, traveling salesman polytopes) and points out algorithmic features (good algorithms, $\mathcal{P}$- and $\mathcal{NP}$-problems, separation, cutting planes). The main subject of

A. Schrijver (1983). Min-max results in combinatorial optimization. *Ibidem,* 439-500

are min-max relations, but a large part consists of applications of such relations to polyhedral combinatorics. Particular attention is given to complete descriptions of polyhedra associated with combinatorial optimization problems. Especially worth mentioning is the description of refinement techniques that yield far reaching generalizations from a few fundamental results. Two papers dealing with a special topic are:

M. Grötschel, M.W. Padberg (1985A). Polyhedral theory. E.L. Lawler, J.K. Lenstra, A.H.G. Rinnooy Kan, D.B. Shmoys (eds.). *The Traveling Salesman Problem,* Wiley, Chichester,

M. Grötschel, M.W. Padberg (1985B). Polyhedral algorithms. *Ibidem.*

This is a series of two papers on polyhedral aspects of the traveling salesman problem. The papers contain almost everything about this subject that is known to date. Broad coverage is given to the use of polyhedral methods in the design of practical LP-based cutting plane algorithms. All polyhedral and algorithmical aspects are treated in general terms and are exemplified by the traveling salesman problem. A somewhat older paper - but still of interest - dealing with the same kind of questions as [Grötschel & Padberg 1985A,B] for another combinatorial optimization problem is

E. Balas, M.W. Padberg (1976). Set partitioning: a survey. *SIAM Rev. 18,* 710-760.

[Grötschel & Padberg 1985B] and [Balas & Padberg 1976] in particular deal with the algorithmic implications of polyhedral combinatorics for special $\mathcal{NP}$-hard problems, i.e. how knowledge of classes of facets for polytopes associated with a hard combinatorial optimization problem can be combined with LP-techniques and branch and bound to obtain practically efficient algorithms. A general description of this kind of approach is given in

M. Grötschel (1982). Approaches to hard combinatorial optimization problems. B. Korte (ed.). *Modern Applied Mathematics: Optimization and Operations Research,* North-Holland, Amsterdam, 437-515,

where the practical use - even of partial linear characterizations of combinatorial polytopes - is demonstrated. On the other hand, the paper

C.H. Papadimitriou (1984). Polytopes and complexity. W.R. Pulleyblank (ed.). *Progress in Combinatorial Optimization,* Academic Press, New York, 295-305

shows that (unless $\mathcal{P} = \mathcal{NP}$) many polyhedral questions about $\mathcal{NP}$-hard problems turn out to be even harder than the original problems. So in a theoretical sense, polyhedral combinatorics does not seem to be a feasible line of attack at $\mathcal{NP}$-hard problems. The word theoretical should be underlined here, since the success of algorithms for $\mathcal{NP}$-hard problems based on results of polyhedral combinatorics shows the contrary empirically. Successful practical experiences with real-world applications of this kind are described in a number of papers, for instance in [Grötschel & Padberg 1985B] and

M. Grötschel, M. Jünger, G. Reinelt (to appear). A cutting plane algorithm for the linear ordering problem. *Oper. Res.*

Since the fundamental work of Gomory, see for instance

R.E. Gomory (1958). Outline of an algorithm for integer solutions to linear programs. *Bull. Amer. Math. Soc. 64,* 275-278,

cutting planes have been considered intensively in integer programming from a theoretical as well as a practical point of view. Illuminating articles about the geometrical background and basic techniques for deriving cutting planes are

V. Chvátal (1973). Edmonds polytopes and a hierarchy of combinatorial problems. *Discrete Math. 4,* 305-337,

and (considering a more general case)

A. Schrijver (1980). On cutting planes. *Ann. Discrete Math. 9,* 291-296.

An introductory article giving a good survey of existing cutting plane methods is

L.A. Wolsey (1979). Cutting plane methods. A.G. Holzmann (ed.). *Operations Research Support Methodology,* Dekker, New York, 441-466.

The field of polyhedral combinatorics has received a new and very stimulating impetus through the invention of the *ellipsoid method*. Here polyhedral results and cutting planes can be used in a novel way. In fact, it turns out that finding an optimum solution to a combinatorial optimization problem in polynomial time is equivalent to finding a cutting plane for the associated polytope in polynomial time. The latter problem is called *separation problem* and asks whether for a given point y and a polyhedron P the fact that y is in P or not can be proved and if $y \notin P$ a hyperplane separating y from P (a cutting plane) can be found. Generalizations of the ellipsoid method and their use together with polyhedral combinatorics for the polynomial-time solvability of combinatorial optimization problems are discussed in

M. Grötschel, L. Lovász, A. Schrijver (1981). The ellipsoid method and its consequences in combinatorial optimization. *Combinatorica 1,* 169-197,

R. M. Karp, C.H. Papadimitriou (1982). On linear characterizations of combinatorial optimization problems. *SIAM J. Comput. 11,* 620-632,

M.W. Padberg, M.R. Rao (to appear). The Russian method for linear inequalities III: bounded integer programming. *Math. Programmming Stud.*

The geometrical flavor of this approach together with an outline of several applications is described in the paper

M. Grötschel, L. Lovász, A. Schrijver (1984). Geometric methods in combinatorial optimization. W.R. Pulleyblank (ed.). *Progress in Combinatorial Optimization,* Academic Press, New York, 167-183,

which was prepared for an instructional series on this subject and avoids the ugly technical details which are necessary for a rigorous treatment of the subject. Such a treatment together with further generalizations and more applicatins will be contained in the forthcoming book

M. Grötschel, L. Lovász, A. Schrijver (to appear). *The Ellipsoid Method and Combinatorial Optimization,* Springer, Berlin.

The application of the ellipsoid method to a combinatorial optimization problem mainly rests on the fact that for the associated polyhedron a complete linear characterization is known. A widely used technique for obtaining such results exploits what is called *total dual integrality.* The importance of this concept for polyhedral combinatorics, algorithms and complexity is described in the instructional paper

J. Edmonds, R. Giles (1984). Total dual integrality of linear inequality systems. W.R. Pulleyblank (ed.). *Progress in Combinatorial Optimization,* Academic Press, New York, 117-129.

Duality plays a central role in polyhedral combinatorics (and not only here). Two special geometric duality theories, called *blocking* and *anti-blocking theory*, have given many new insights into the relations between various combinatorial problems - in particular for packing and covering type problems. A seminal paper on this subject is

D.R. Fulkerson (1971). Blocking and anti-blocking pairs of polyhedra. *Math. Programming 1,* 168-194.

It turned out that the investigation of anti-blocking relations can be viewed as the study of certain problems on perfect graphs. A beautiful theory with many deep results has evolved here. The book

C. Berge, V. Chvátal (eds.) (1984). *Topics on Perfect Graphs, Ann. Discrete Math. 21*

contains most of the fundamental papers on this subject and a number of new results. So a very good overview over the historical development of this area up to very recent research can be gained from this book.

A subject that has to be mentioned here as well is the work of Seymour on general *min-max relations.* Seymour has extended a number of classic min-max relations like various versions of the max-flow min-cut theorem to quite general frameworks. He states his results usually in matroid language and often characterizes the matroids for which certain min-max relations hold via forbidden minor theorems. His work has quite a polyhedral flavor, since polyhedral results (e.g. LP-duality or complementary slackness) are used quite often and many of his theorems can be nicely interpreted in the language of polyhedral combinatorics. Two papers we would like to mention in this area are

P.D. Seymour (1977). The matroids with the max-flow min-cut property. *J. Comb. Theory Ser. B 23,* 189-222,

P.D. Seymour (1981). Matroids and multicommodity flows. *European J. Combin. 2,* 257-290.

These articles are not easy to read but worth the effort.

A subject that grew out of the study of matroid polytopes and polymatroids is the investigation of *submodular functions*; see [Edmonds 1970] above. These functions are in some sense the combinatorial analogues of convex functions and there are a number of interesting polyhedra, separation theorems and min-max relations associated with them. A survey of this topic is given in

L. Lovász (1983). Submodular functions and convexity. A. Bachem, M. Grötschel, B. Korte (eds.). *Mathematical Programming: the State of the Art - Bonn 1982*, Springer, Berlin, 235-257.

Another line of attack at combinatorial problems from a polyhedral point of view is based on Gomory's group problem and his theory of corner polyhedra. More generally, it is now often called the *subadditive approach* to integer programming. The theory of corner polyhedra can be found in

A. Bachem (1976). *Beiträge zur Theorie der Corner Polyeder,* Mathematical

Systems in Economics 29, Hain, Meisenheim am Glan,

and a survey of subadditive techniques in

E.L. Johnson (1979). On the group problem and a subadditive approach to integer programming. *Ann. Discrete Math. 5*, 97-112.

A recent research paper showing nice applications of this approach is

G. Gastou, E.L. Johnson (to appear). Binary group and Chinese postman polyhedra. *Math. Programming.*

Polyhedral combinatorics does of course have its special branches. Flourishing areas are for instance the hunt for *facets* of polyhedra associated with combinatorial problems (see the papers under the subject 'facets of integer polyhedra' in [Kastning 1976], [Hausmann 1978], [Von Randow 1982], or see [Pulleyblank 1983] and [Grötschel & Padberg 1985A]) and the hunt for *complete linear characterizations* of such polyhedra (see [Pulleyblank 1983] and [Schrijver 1983]). Another topic of interest is the study of *adjacency relations* on combinatorial polytopes. A good treatment of this subject is

D. Hausmann (1980). *Adjacency on Polytopes in Combinatorial Optimization,* Mathematical Systems in Economics 49, Hain, Meisenheim am Glan.

There are a number of *expository papers* that describe certain polyhedral or linear programming aspects of combinatorial problems. We want to mention a few of them. The paper

V. Chvátal (1975). Some linear programming aspects of combinatorics. D. Corneil, E. Mendelsohn (eds.). *Proc. Conf. Algebraic Aspects of Combinatorics, Toronto, 1975,* Congressus Numerantium 13, Utilitas Mathematica, Winnipeg, 2-30

describes basic examples and notions and is aimed at an audience with no previous knowledge of linear programming. The exposition is very nice and worth reading.

A.J. Hoffman (1982). Ordered sets and linear programming. I. Rival (ed.). *Ordered Sets,* Reidel, Dordrecht, 619-634

shows how linear programming ideas can be used to prove theorems about ordered sets (Dilworth's theorem and generalizations such as the theorems of Greene and Greene & Kleitman) and, the other way around, how concepts from partially ordered sets can be used in polyhedral theory (study of lattice polyhedra).

Various roles of *totally unimodular matrices* (these are matrices for which each square submatrix has determinant 1, 0 or -1) in polyhedral combinatorics are surveyed in

A.J. Hoffman (1960). Some recent applications of the theory of linear inequalities to extremal combinatorial analysis. R. Bellman, M. Hall (eds.).

Combinatorial Analysis, American Mathematical Society, Providence, RI, 113-128,

A.J. Hoffman (1976). Total unimodularity and combinatorial theorems. *Linear Algebra Appl. 13,* 103-108,

A.J. Hoffman (1979). The role of unimodularity in applying linear inequalities to combinatorial theorems. *Ann. Discrete Math. 4,* 73-84.

In [Hoffman 1960] emphasis is laid on Hall's theorem, several variations and generalizations are stated and its relations to transportation and transshipment problems is explored. [Hoffman 1976] shows how total unimodularity can be used to give a common proof to a variety of transversal theorems, and [Hoffman 1979] states a number of interesting results about polytopes associated with combinatorial optimization problems.

We have already mentioned before that *duality* plays a central role in many aspects of combinatorial optimization. A large number of results of this type are surveyed in

L. Lovász (1977). Certain duality principles in integer programming. *Ann. Discrete Math. 1,* 363-374.

An overview of polytopes associated with combinatorial optimization problems can be found in

L. Lovász (1979). Graph theory and integer programming. *Ann. Discrete Math. 4,* 141-158.

Here also several examples are presented illuminating some of the proof techniques used in polyhedral combinatorics.

As mentioned above, there exist no monographs (yet) covering a large part of the field 'polyhedral combinatorics', but there are a number of proceedings volumes or paper collections that mainly contain articles on this subject. To mention some of these books which are worth looking at to get an idea about the present research topics we quote:

P.L. Hammer, E.L. Johnson, B.H. Korte, G.L. Nemhauser (eds.) (1977). *Studies in Integer Programming, Ann. Discrete Math. 1,*

M.L. Balinski, A.J. Hoffman (eds.) (1978). *Polyhedral Combinatorics, Math. Programming Stud. 8,*

P.L. Hammer, E.L. Johnson, B.H. Korte (eds.) (1979). *Discrete Optimization I, Ann. Discrete Math. 4,*

P.L. Hammer, E.L. Johnson, B.H. Korte (eds.) (1979). *Discrete Optimization II, Ann. Discrete Math. 5,*

M.W. Padberg (ed.) (1980). *Combinatorial Optimization, Math. Programming Stud. 12,*

A. Bachem, M. Grötschel, B. Korte (eds.) (1982). *Bonn Workshop on*

Combinatorial Optimization, Ann. Discrete Math. 16.

This survey of the literature on polyhedral combinatorics is by no means complete, but we think it gives a reasonable collection of papers and books which may be suitable for the beginner to start exploring the subject. We have mainly concentrated on very recent books and papers some of which have not appeared yet but are circulating as preprints. Most of these papers will however appear in the years 1984 and 1985.

2

Duality for Integer Optimization

G.L. Nemhauser
Cornell University, Ithaca

Duality is fundamental to the theory, algorithms and post-optimality analysis of linear programming. We assume that the reader is familar with this subject. Any good textbook on linear programming can be used as a reference; see, for example,

G.B Dantzig (1963). *Linear Programming and Extensions,* Princeton University Press, Princeton, NJ,

V. Chvátal (1983). *Linear Programming,* Freeman, New York.

Duality theory for integer optimization is in a much less settled state; nevertheless, it is immensely important to the subject.

Several of the papers to be cited here are contained in three recent books that are collections of survey papers. These are

P.L. Hammer, E.L. Johnson, B.H. Korte (eds.) (1979). *Discrete Optimization I, Ann. Discrete Math. 4,*

P.L. Hammer, E.L. Johnson, B.H. Korte (eds.) (1979). *Discrete Optimization II, Ann. Discrete Math. 5,*

A. Bachem, M. Grötschel, B. Korte (eds.) (1983). *Mathematical Programming: the State of the Art - Bonn 1982,* Springer, Berlin.

In addition, two papers in this volume are closely related to the subject of this bibliography and contain many overlapping references. These are

M. Grötschel (1985). Polyhedral combinatorics. *This volume,* Ch. 2,

L.E. Trotter, Jr. (1985). Discrete packing and covering. *This volume,* Ch. 4.

The principal result of duality is a max-min relationship between a pair of optimization problems. Suppose the *primal* problem is

$$P: \max\{z(x): x \in S\}.$$

The problem

$$D: \min\{w(y): y \in T\}$$

is said to be *weakly dual* to P if for all $x \in S$ and $y \in T$, we have $w(y) \leqslant z(x)$. Weak duality implies that whenever P has an optimal solution

$$Z^P = \max\{z(x): x \in S\} \leqslant w(y) \quad \text{for all } y \in T.$$

The least upper bound on the optimal value of P that can be obtained from a feasible solution to D is given by

$$W^D = \min\{w(y): y \in T\}.$$

The *duality gap* between P and D is defined to be $W^D - Z^P$.

When the duality gap is zero, i.e. $W^D = Z^P$, D is said to be *strongly dual* to P. Whenever we have a pair (x,y) of primal and dual feasible solutions with the property that $z(x) = w(y)$, then x is an optimal solution to P and y is an optimal solution to D.

Thus strong duality provides a mechanism, essentially the only tool besides enumeration, for proving the optimality of a primal feasible solution. Weak duality may yield a good bound that can be used to prove that a primal feasible solution is nearly optimal and such bounds are also essential to curtailing the amount of enumeratiom required in a branch-and-bound algorithm.

When $z(x) = cx$ and $S = \{x: Ax \leqslant b, x \text{ integral}\}$, the primal problem is the integer program

$$(IP) \quad \max\{cx: Ax \leqslant b, x \text{ integral}\}.$$

It is well known that (IP) is equivalent to the linear program

$$(LP) \quad \max\{cx: x \in conv(S)\}$$

where $conv(S)$ is the convex hull of the points in S. Hence one might be tempted to dismiss integer programming duality, and for that matter the whole subject of integer programming, as being nothing more than linear programming. But this superficial viewpoint is not productive. There are significant distinctions to be made depending on the structure of $conv(S)$.

Polyhedral combinatorics focuses on those situations in which $conv(S)$ has a nice description. Since the literature of this subject is amply covered in the above mentioned bibliographies of Grötschel and Trotter, our treatment of it will be brief. An excellent basic reference is the recent survey paper

W.R. Pulleyblank (1983). Polyhedral combinatorics. [Bachem, Grötschel & Korte 1983], 312-345.

The corresponding duality results are surveyed in

A. Schrijver (1983). Min-max results in combinatorial optimization. [Bachem, Grötschel & Korte 1983], 439-500.

Both papers are highly recommended for their fine exposition and comprehensive coverage.

Some earlier relevant surveys of topics in polyhedral combinatorics are

C. Berge (1979). Packing problems and hypergraph theory: a survey. *Ann. Discrete Math. 4,* 3-38,

A.J. Hoffman (1979). The role of unimodularity in applying linear inequalities to combinatorial theorems. *Ann. Discrete Math. 4,* 73-84,

L. Lovász (1979). Graph theory and integer programming. *Ann. Discrete Math. 4,* 141-158,

J. Tind (1979). Blocking and antiblocking polyhedra. *Ann. Discrete Math. 4,* 159-174,

M.W. Padberg (1979). Covering, packing and knapsack problems. *Ann. Discrete Math. 4,* 265-288,

D.R. Fulkerson (1971). Blocking and anti-blocking pairs of polyhedra. *Math. Programming 1,* 168-194.

The simplest case of a nice description of $conv(S)$ occurs when matrix A is *totally unimodular.* Here $conv(S) = \{x : Ax \leqslant b\}$ so that strong duality in integers is a direct consequence of linear programming duality. So, for example, we get the famous max-flow min-cut duality theorem and many other combinatorial duality results associated with network flows; see the classical book

L.R. Ford, D.R. Fulkerson (1962). *Flows in Networks,* Princeton University Press, Princeton, NJ.

A more recent treatment of this subject is given in

E.L. Lawler (1976). *Combinatorial Optimization: Networks and Matroids,* Holt, Rinehart and Winston, New York,

Given the importance of unimodularity in the development of combinatorial optimization, it is rather surprising that the problem of polynomial-time recognition of a totally unimodular matrix and the problem of solving totally unimodular linear programs in polynomial time have been resolved only recently; see

P.D. Seymour (1980). Decomposition of regular matroids. *J. Combin. Theory Ser. B 28,* 305-359,

R.G. Bland, J. Edmonds (1982). *Fast Primal Algorithms for Totally Unimodular Linear Programming,* Presented at the XI International Symposium on Mathematical Programming, Bonn, F.R.G.

The first, and perhaps most famous, instance of nontrivial integer programming strong duality concerns the matching problem; see

J. Edmonds (1965). Paths, trees, and flowers. *Canad. J. Math. 17,* 449-467,

J. Edmonds (1965). Maximum matching and a polyhedron with 0-1 vertices. *J. Res. Nat. Bur. Standards 69B,* 125-130.

A matching in a graph is a collection of edges no two of which meet at a common vertex. Edmonds gave a polynomial-time algorithm for the problem of finding a maximum weight matching. The algorithm actually solves a linear program over a polytope S_M that clearly contains the convex hull of matchings. However, for any objective function, Edmonds proved that the algorithm produces an optimal primal solution which is a matching, i.e. an integral solution. This proves that S_M is the convex hull of matchings and also establishes a strong duality between weighted matchings and a class of objects that cover the weighted edges of a graph. Thus we simultaneously get a polynomial-time algorithm, a polyhedral characterization and a strong duality for matchings. In fact, for good reasons, it has turned out to be the rule, rather than the exception, for these three types of results to occur together.

Edmonds' pioneering work stimulated many other results of this type for combinatorial optimization problems, especially those associated with packing and covering problems on matroids and graphs. We refer the interested reader to the surveys and bibliographies cited above, particularly [Grötschel 1985], [Trotter 1985], [Pulleyblank 1983], [Schrijver 1983] above.

We close this part of the bibliography by mentioning another duality result that is true for matchings. A polyhedron given by the constraints $Ax \leqslant b$ is called *totally dual integral* if for each integral, primal objective function c for which the linear programming dual problem $\min\{yb: yA = c, y \geqslant 0\}$ has an optimal solution, it has an integral optimal solution y. The fundamental result proved in

J. Edmonds, R. Giles (1977). A min-max relation for submodular functions on graphs. *Ann. Discrete Math. 1*, 185-205

is that if $Ax \leqslant b$ is totally dual integral, then for each integral b such that $\max\{cx: Ax \leqslant b\}$ has an optimal solution, it has an integral optimal solution. An important consequence of total dual integrality is that we obtain a purely combinatorial max-min relationship; i.e., both the primal and dual linear programming problems have integral solution vectors. Several totally dual integral systems are known, see [Trotter 1985], [Pulleyblank 1983] and especially [Schrijver 1983] above.

So far we have discussed integer programs where an explicit linear programming duality is known. One step removed are integer programs with the integer-rounding property. Let

$$z^o_{IP}(c) = \max\{cx: Ax \leqslant 1, x \geqslant 0 \text{ and integral}\},$$

$$z^o_{LP}(c) = \max\{cx: Ax \leqslant 1, x \geqslant 0\}.$$

The family of integer programs generated by all objective functions c is said to have the *integer-rounding* property if for all c, $z^o_{LP}(c) - z^o_{IP}(c) < 1$. When the integer-rounding property holds, the dual of the linear programming relaxation provides a weak dual to the integer program with the property that the difference between the dual and primal optimal objective values is always less than 1. Trotter's bibliography covers this topic in detail.

Since any integer program can in principle be reformulated as a linear program, it has a strong dual which is a linear programming problem. A different strong duality for general integer programming problems will be introduced later. Nevertheless, for most integer programming problems of practical interest, i.e. those that are $\mathcal{NP}$-hard, we do not know a strong dual program that would be useful computationally in producing a polynomial-time algorithm. In fact there is evidence from the theory of computational complexity that no such nice descriptions exist. The complexity aspects of this subject are studied in

R.M. Karp, C.H. Papadimitriou (1982). On linear characterizations of combinatorial optimization problems. *SIAM J. Comput. 11,* 620-632,

C.H. Papadimitriou (1984). Polytopes and complexity. W.R. Pulleyblank (ed.). *Progress in Combinatorial Optimization,* Academic Press, New York, 295-305.

Fortunately, weak duality imbedded into a branch-and-bound algorithm can provide reasonably efficient algorithms for a variety of $\mathcal{NP}$-hard integer programming problem. Notable among these approaches is *Lagrangean duality.* Two historically important papers on the use of Lagrange multipliers in integer programming are

J. Lorie, L. Savage (1955). Three problems in capital rationing. *J. Business 28,* 229-239,

H. Everett (1963). Generalized Lagrange multiplier method for solving problems of optimum allocation of resources. *Oper. Res. 11,* 399-417.

Consider the integer program $\max\{cx : Ax \leqslant b, x \in S\}$. Assume that it is relatively easy to maximize a linear function over S, but that the constraints $Ax \leqslant b$ make the problem much more difficult. For example, S could be the constraint set of a network flow problem. Lagrangean duality capitalizes on this structure. In particular, the Lagrangean dual of the integer program is

$$\min\{z(\lambda): \lambda \geqslant 0\}, \text{ where } z(\lambda) = \max\{(c - \lambda A)x + \lambda b : x \in S\}.$$

The duality result here is

$$\min\{z(\lambda): \lambda \geqslant 0\} = \max\{cx : Ax \leqslant b, x \in conv(S)\}.$$

This duality is weak because

$$\max\{cx : Ax \leqslant b, x \in S\} \leqslant \max\{cx : Ax \leqslant b, x \in conv(S)\}$$

and it is generally not true that the inequality holds as an equality. The function $z(\lambda)$ is piecewise linear and convex, and subgradient optimization has proved to be a useful technique for optimizing it, see

M. Held, P. Wolfe, H.D. Crowder (1974). Validation of subgradient optimization. *Math. Programming 6,* 62-88.

The current use of Lagrangean duality for solving structured $\mathcal{NP}$-hard combinatorial optimization problems was inspired by the papers

M. Held, R.M. Karp (1970). The traveling-salesman problem and minimum spanning trees. *Oper. Res. 18,* 1138-1162,

M. Held, R.M. Karp (1971). The traveling-salesman problem and minimum spanning trees: Part II. *Math. Programming 1,* 6-25.

Held and Karp used Lagrangean duality and branch-and-bound to solve larger traveling salesman problems than could be dealt with by earlier approaches. A different Lagrangean dual approach to the traveling salesman problem is given in

E. Balas, N. Christofides (1981). A restricted Lagrangian approach to the traveling salesman problem. *Math. Programming 21,* 19-46.

A very clear exposition and formalization of the use of Lagrangean duality in integer programming is the paper

A.M. Geoffrion (1974). Lagrangian relaxation for integer programming. *Math. Programming Stud. 2,* 82-114.

Lagrangean duality has been used to solve problems in scheduling, location, distribution and even general pure integer programs in combination with the group-theoretic approach. Two main contributors in this area have written excellent survey papers:

J.F. Shapiro (1979). A survey of Lagrangian techniques for discrete optimization. *Ann. Discrete Math. 5,* 113-138,

M.L. Fisher (1981). The Lagrangian relaxation method for solving integer programming problems. *Management Sci. 27,* 1-18.

The key to solving large $\mathcal{NP}$-hard combinatorial optimization problems efficiently by branch-and-bound is to produce a formulation for which the duality gap will be small so that the potentially exponential time enumerative phase of the algorithm will not explode. Lagrangean duality is one instance of this approach.

Curiously, and almost mysteriously, there are some problems for which the duality gap can be large, but for most data sets the gap is small or even zero. This phenomenon occurs in the uncapacitated facility location problem; see

D. Erlenkotter (1978). A dual-based procedure for uncapacitated facility location. *Oper. Res. 26,* 992-1009,

J. Krarup, P.M. Pruzan (1983). The simple plant location problem: survey and synthesis. *European J. Oper. Res. 12,* 36-81,

G. Cornuéjols, G.L. Nemhauser, L.A. Wolsey (to appear). The uncapacitated facility location problem. R.L. Francis, P. Mirchandani (eds.). *Discrete Location Theory,* Wiley, New York.

Erlenkotter showed empirically that simple heuristics for both the primal and dual problems sufficed to prove optimality in almost all of the instances he dealt with and in the remaining instances the heuristics produced nearly optimal solutions so that the solution could be completed with a modest amount of enumeration. This idea has been pursued successfully in the optimization of a complex distribution system:

M.L. Fisher, A.J. Greenfield, R. Jaikumar, P. Kedia (1982). *Real-Time Scheduling of a Bulk Delivery Fleet: Practical Application of Lagrangian Relaxation,* Report 82-10-11, The Wharton School, University of Pennsylvania, Philadelphia,

and appears to be a promising approach for obtaining good solutions to practical problems.

While the usefulness of Erlenkotter's heuristics for the uncapacitated facility location problem has only been established empirically, sometimes weak duality can be used to establish provable worst-case bounds on the quality of heuristic solutions. This idea has been exposited in

L.A. Wolsey (1980). Heuristic analysis, linear programming and branch-and-bound. *Math. Programming Stud. 13,* 121-134,

where applications in the literature to bin packing, longest Hamiltonian cycle, Euclidean traveling salesman, set covering and location problems are discussed.

One interpretation of *cutting planes* for integer programs is that they strenghten dual formulations. Here we present this point of view. The linear inequality $ax \leqslant b$ is *valid* for the set Q if $Q \subseteq \{x: ax \leqslant b\}$. Suppose $S^i \supseteq S$ is a polyhedron. Then the linear programming dual of $\max\{cx: x \in S^i\}$ is weakly dual to the integer program $\max\{cx: x \in S\}$. Now if $ax \leqslant b$ is valid for S but not for S^i, a stronger dual formulation is the linear programming dual of

$$\max\{cx: x \in S^{i+1}\} \text{ where } S^{i+1} = S^i \cap \{x: ax \leqslant b\}.$$

In cutting plane algorithms, one begins with a polyhedron $S^0 \supset S$ with the property that if $x \in S^0$ and integral, then $x \in S$. Hence strong duality is obtained for the ith subproblem if its solution vector x^i is integral. If x^i is not integral, we determine a valid inequality for S that separates x^i from S^i, i.e. $x^i \notin S^{i+1}$. However, the optimal dual solution to the ith subproblem is feasible to the dual of the $(i+1)$st subproblem, which can then be solved by the dual simplex algorithm.

This cutting plane approach was first proposed in two classical papers on the traveling salesman problem:

G.B. Dantzig, D.R. Fulkerson, S.M. Johnson (1954). Solution of a large scale travelling salesman problem. *Oper. Res. 2,* 393-410,

G.B. Dantzig, D.R. Fulkerson, S.M. Johnson (1959). On a linear programming, combinatorial approach to the travelling salesman problem. *Oper. Res. 7,* 58-66.

A general, finite cutting plane algorithm for integer programs was developed in

R.E. Gomory (1963). An algorithm for integer solutions to linear programs. R.L. Graves, P. Wolfe (eds.). *Recent Advances in Mathematical Programming,* McGraw-Hill, New York, 269-302.

A simple characterization of all valid inequalities for an integer program is given in

V. Chvátal (1973). Edmonds polytopes and a hierarchy of combinatorial problems. *Discrete Math. 4,* 305-317.

The fundamental theory of cutting plane methods is exposited in

R. Jeroslow (1979). An introduction to the theory of cutting-planes. *Ann. Discrete Math. 5,* 71-95,

and a more comprehensive treatment of this material appears in

R. Jeroslow (1978). Cutting plane theory: algebraic methods. *Discrete Math. 23,* 121-150.

Gomory's cutting plane algorithm proved to be too slow to solve significant integer programming problems. While his valid inequalities are generated rapidly, the monotonically decreasing dual bounds cx^i, $i = 0,1,...$, converge too slowly to the optimal value of the primal objective function. For cutting plane algorithms to be practical, strong valid inequalities and, in particular, facets of $conv(S)$ are needed. This point was clearly demonstrated in the papers

M. Grötschel (1980). On the symmetric travelling salesman problem: solution of a 120 city problem. *Math. Programming Stud. 12,* 61-77,

M.W. Padberg, S. Hong (1980). On the symmetric travelling salesman problem: a computational study. *Math. Programming Stud. 12,* 78-107,

where facets of the traveling salesman polytope are generated and used as cutting planes in the manner described above.

These papers have inspired the development of new cutting plane algorithms that use strong valid inequalities derived from the structure of the problem. An algorithm for the general 0-1 integer programming problem is given in

H. Crowder, E.L. Johnson, M.W. Padberg (1983). Solving large-scale zero-one linear programming problems. *Oper. Res. 31,* 803-834.

Fixed charge problems are studied from this point of view in

M.W. Padberg, T.J. Van Roy, L.A. Wolsey (1982). *Valid Inequalities for Fixed Charge Problems,* CORE Discussion paper 8232, Université Catholique de Louvain,

and some other mixed integer programs in

T.J. Van Roy, L.A. Wolsey (1983). *Valid Inequalities for Mixed 0-1 Programs,* CORE Discussion paper 8316, Université Catholique de Louvain.

Much work remains to be done in this fruitful area of research.

Gomory's work on cutting plane algorithms, his work with Gilmore on the knapsack problem:

P.C. Gilmore, R.E. Gomory (1966). The theory and computation of knapsack functions. *Oper. Res. 14,* 1045-1074,

and his work on the group problem of integer programming:

R.E. Gomory (1969). Some polyhedra related to combinatorial problems. *Linear Algebra Appl. 2,* 451-558,

have led to a superadditive cutting plane theory and duality theory for integer programs. Principal contributors have been Ellis Johnson and Robert Jeroslow, see the expository papers [Jeroslow 1979] above and

E.L. Johnson (1979). On the group problem and a subadditive approach to integer programming. *Ann. Discrete Math. 5,* 97-112.

A real-valued function f is *superadditive* on $D \subseteq R^m$ if $f(a)+f(b) \leqslant f(a+b)$ for all $a, b, a+b \in D$. There are two fundamental results concerning superadditive functions and the integer programming problem

$$\max\{\sum_{j=1}^{n} c_j x_j : \sum_{j=1}^{n} a_j x_j \leqslant b,\ x_j \geqslant 0 \text{ and integer},\ j = 1,\ldots,n\};$$

see [Jeroslow 1978] and [Jeroslow 1979] above. The problem

$$\min\{F(b):\ F(a_j) \geqslant c_j,\ j = 1,\ldots,n,\ F(0) = 0,\ F \text{ superadditive and nondecreasing}\}$$

is strongly dual to the integer program. A closely related result is that if $F(0) = 0$ and F is superadditive and nondecreasing, then

$$\sum_{j=1}^{n} F(a_j) x_j \leqslant F(b)$$

is a valid inequality for the constraint set of the integer program; moreover all maximal valid inequalities can be generated by such functions.

A solution to the dual is the value function of the integer program, i.e.,

$$F(b) = \max\{\sum_{j=1}^{n} c_j x_j : \sum_{j=1}^{n} a_j x_j \leqslant b,\ x_j \geqslant 0 \text{ and integer},\ j = 1,\ldots,n\},$$

with appropriate consideration of values of b that yield either primal infeasibility or unboundedness. But this doesn't tell us very much without a sharper characterization of the value function. A recursive description of the value function is given in

C.E. Blair, R.G. Jeroslow (1982). The value function of an integer program.

Math. Programming 23, 237-273;

see also

L.A. Wolsey (1981). The b-hull of an integer program. *Discrete Appl. Math. 3,* 193-201.

The superadditive approach has been extended to mixed integer programs. Two basic papers are

E.L. Johnson (1974). The group problem for mixed integer programming. *Math. Programming Stud. 2,* 137-179,

C.E. Blair, R.G. Jeroslow (1981). *Constructive Characterization of the Value Function of a Mixed-Integer Program,* BEBR Working paper 784, University of Illinois.

Some of Johnson's work has focused on dual algorithms that construct superadditive functions, see

C.A. Burdet, E.L. Johnson (1977). A subadditive approach to solve integer programs. *Ann. Discrete Math. 1,* 117-144,

E.L. Johnson (1980). Subadditive lifting methods for partitioning and knapsack problems. *J. Algorithms 1,* 75-96.

A very readable presentation concerning the different types of dual functions constructed by various integer programming algorithms is given in

L.A. Wolsey (1981). Integer progamming duality: price functions and sensitivity analysis. *Math. Programming 20,* 173-195.

Wolsey interprets the dual functions as prices and shows how they can be used in sensitivity analysis.

3

Discrete Packing and Covering

L.E. Trotter, Jr.
Cornell University, Ithaca

CONTENTS

1. INTRODUCTION

Suppose $G = (V,E)$ is a digraph with vertices V, edges E and with distinct $s,t \in V$. We denote by $A = [a_{ij}]$ the (edge-)incidence matrix of directed (s,t)-paths in G, whose rows index the (s,t)-dipaths of G and whose columns index the edges of G; thus $a_{ij} = 1$ when the jth edge appears in the ith (s,t)-dipath and otherwise $a_{ij} = 0$. Then for any nonnegative vector of edge capacities $c = (c(e):e \in E)$ the maximum flow problem can be modeled as

$$P(A,c)\text{: } \max\{1\cdot y : yA \leqslant c, y \geqslant 0\}.$$

In the introductory pages of

L.R. Ford, Jr., D.R. Fulkerson (1962). *Flows in Networks,* Princeton University Press, Princeton, NJ,

this path-edge formulation of the maximum flow problem is shown to be equivalent to the 'standard' formulation based on the vertex-edge incidence matrix of G.

In the integer programming restriction of $P(A,c)$,

$$P_I(A,c)\text{: } \max\{1\cdot y : yA \leqslant c, y \geqslant 0, y \text{ integral}\},$$

one seeks a maximum cardinality (integral) 'packing' of the rows of A into the vector c. For any nonnegative matrix A and any nonnegative vector c we will refer to $P_I(A,c)$ as a *discrete packing model.* Thus the integral maximum flow problem with integral capacities is termed an (s,t)-dipath packing model. For discrete packing models it is generally the case that A is an integral matrix and one usually insures boundedness of objective value in $P_I(A,c)$ by assuming that A has no row all of whose entries are 0.

Also for any nonnegative A and c, the problem

$$C_I(A,c)\text{: } \min\{1 \cdot y : yA \geqslant c, y \geqslant 0, y \text{ integral}\}$$

is termed a *discrete covering model.* Consider, for instance, the case in which A is the (vertex-)incidence matrix of stable sets in a simple graph G; $C_I(A,c)$ is then then weighted vertex coloring problem for G. In most covering applications A will again be an integral matrix and usually feasibility of $C_I(A,c)$ for all nonnegative vectors c is guaranteed by assuming that A has no 0-valued columns. The linear programming relaxation of $C_I(A,c)$, denoted $C_I(A,c)$, is obtained by deleting the integrality stipulation in $C_I(A,c)$.

In the sequel we categorize and briefly survey certain discrete packing and covering models. The models which we consider will be well-behaved in the sense that for a given matrix A the optimum *values* of $P(A,c)$ and $P_I(A,c)$, or the analogous *values* for covering models, will always be equal or nearly equal, i.e., for all nonnegative integral c. For brevity we omit explicit definitions of the combinatorial structures which give rise to these models; for such definitions the reader is referred to the references cited. For additional surveys of this and closely related topics, the reader should consult the six references below, which we now briefly describe. [Fulkerson 1971] provides the initial and fundamental survey of blocking theory and antiblocking theory. These are polyhedral duality theories treating, respectively, polyhedra of the forms given by the linear programming duals of $P(A,c)$ and $C(A,c)$. Important initial examples of discrete packing and covering models are presented here. In Chapter 2 of [Baum 1977] several combinatorial optimization models are discussed in a manner stressing their common algebraic features and a general survey of discrete packing and covering models is presented in Chapter 2 of [Marcotte 1983]. [Grötschel, Lovász & Schrijver 1981] demonstrate that the ellipsoid method provides an important and powerful tool for the analysis of discrete packing and covering models. Finally, in [Schrijver 1983] combinatorial min-max statements arising from discrete packing and covering models are surveyed and in [Schrijver 1984] min-max statements are studied from the viewpoint of establishing such results using total dual integrality (see §2 below) arguments. These comprehensive surveys provide an indispensible overview of discrete packing and covering models.

D.R. Fulkerson (1971). Blocking and anti-blocking pairs of polyhedra. *Math. Programming 1,* 168-194.

S. Baum (1977). *Integral Near-Optimal Solutions to Certain Classes of Linear Programming Problems,* Technical report 360, School of Operations Research and Industrial Engineering, Cornell University, Ithaca, NY.

O.M.-C. Marcotte (1983). *Topics in Combinatorial Packing and Covering,* Ph.D. thesis, School of Operations Research and Industrial Engineering, Cornell University, Ithaca, New York.

M. Grötschel, L. Lovász, A. Schrijver (1981). The ellipsoid method and its consequences in combinatorial optimization. *Combinatorica 1,* 169-197.

A. Schrijver (1983). Min-max results in combinatorial optimization. A. Bachem, M. Grötschel, B. Korte (eds.). *Mathematical Programming: the State of the Art - Bonn 1982,* Springer, Berlin, 439-500.

A. Schrijver (1984). Total dual integrality from directed graphs, crossing families, and sub- and supermodular functions. W.R. Pulleyblank (ed.). *Progress in Combinatorial Optimization,* Academic Press, New York, 315-361.

2. Strong Integrality

Considering again the maximum flow problem we note that generally the dipath incidence matrix will not be totally unimodular. Nevertheless, when A is any dipath incidence matrix the following *strong max-min property* (see [Fulkerson 1971], §1) is valid: for any nonnegative integral vector c, $P(A,c)$ has an integer-valued optimum solution vector. This is not difficult to show by interpreting $P(A,c)$ and its linear programming dual in light of the max-flow min-cut theorem (see [Ford & Fulkerson 1962], §1). Analogously, when A is the (edge-)incidence matrix of vertex stars in a bipartite graph, it follows from the famous theorem of König (min vertex cover = max matching) that the following *strong min-max property* (defined in [Fulkerson 1971]) holds: $C(A,c)$ and $C_I(A,c)$ have the same optimum solution value for any nonnegative integral vector c. The terminology used here stems from the fact that the strong max-min and min-max properties give rise to discrete or combinatorial strengthenings of the usual max-min and min-max statements arising from linear programming duality considerations. In the present section we indicate several combinatorial instances for which such *strong integrality* results hold.

As noted above, the maximum flow problem provides a prototypical discrete packing model for which strong integrality holds. It follows from

J. Edmonds (1972). Edge-disjoint branchings. R. Rustin (ed.). *Combinatorial Algorithms,* Algorithmics Press, New York, 91-96,

also an early and fundamental result in this area, that strong integrality holds for the incidence matrix of rooted spanning branchings in a digraph. For both of these examples, blocking duality (see [Fulkerson 1971]) suggests a 'dual' instance of strong integrality. For the maximum flow case, the related family of positive parts of minimal (s,t)-cuts gives a discrete packing model for which strong integrality holds - see [Fulkerson 1971] and

J.T. Robacker (1956). *Min-Max Theorems on Shortest Chains and Disjunct Cuts of a Network,* Research memorandum RM-1660-PR, RAND Corporation, Santa Monica, CA.

With regard to rooted spanning branchings, strong integrality for positive parts of rooted cuts follows from

J. Edmonds (1968). Optimum branchings. G.B. Dantzig, A.F. Veinott, Jr. (eds.). *Mathematics of the Decision Sciences,* Lectures in Applied Mathematics 11, American Mathematical Society, Providence, RI, 346-361,

D.R. Fulkerson (1974). Packing rooted directed cuts in a weighted directed graph. *Math. Programming 6,* 1-14.

Note that the previous two examples arise from classes of cutsets in a digraph. A related, though apparently deeper, result of Lucchesi and Younger establishes strong integrality for the (edge-)incidence matrix of directed cutsets in a digraph:

C.L. Lucchesi, D.H. Younger (1978). A minmax relation for directed graphs. *J. London Math. Soc. (2) 17,* 369-374.

In contrast to the earlier examples, it has been shown by Schrijver in

A. Schrijver (1980). A counterexample to a conjecture of Edmonds and Giles. *Discrete Math. 32,* 213-214

that strong integrality *does not* generally hold for the blocking clutter (see [Fulkerson 1971]) of directed cutsets. For particular classes of digraphs, however, for which the latter model *does* have the strong integrality property, see

A. Frank (1979). Kernel systems of directed graphs. *Acta Sci. Math. (Szeged) 41,* 63-76,

A. Schrijver (1982). Min-max relations for directed graphs. *Ann. Discrete Math. 16,* 261-280.

Several recent papers have considered generalizing certain of the above models in a manner so that strong integrality will still hold. Most notably, in

P.D. Seymour (1977). The matroids with the max-flow min-cut property. *J. Combin. Theory Ser. B 23,* 189-222,

a natural matroid generalization of the maximum flow model is considered and a forbidden minor characterization is given for the class of matroids for which the associated discrete packing model has the strong integrality property. Relying on this work of Seymour, Korach has given a characterization of those instances of $P(A,c)$ for which strong integrality will hold when A is the (edge)-incidence matrix of T-cuts in an undirected graph:

E. Korach (1982). *Packings of T-Cuts, and Other Aspects of Dual Integrality,* Ph.D. thesis, Department of Combinatorics and Optimization, University of Waterloo, Ontario.

In [Frank 1979] (see above) Frank introduces the notion of a kernel system of a digraph and uses this combinatorial structure to generalize the maximum flow and rooted spanning branching models mentioned above. A generalization is also obtained for the respective blocking models, namely, the positive parts of minimal (s,t)-cuts and the positive parts of rooted cuts. In [Schrijver 1982] (see above) Schrijver also has generalized these results using the concept of strong connectors for a digraph; the strong integrality result from [Lucchesi & Younger 1978] (see above) can also be deduced using the model of [Schrijver 1982] - see [Schrijver 1984] (§1). The reader is especially referred to [Schrijver 1984] in which the interrelationships among various combinatorial models, including many which give rise to strong integrality results, are detailed. Finally, in

A. Schrijver (to appear). Packing and covering of crossing families of cuts. *J. Combin. Theory Ser. B.*

Schrijver characterizes certain crossing families which, when defined on the vertices of any digraph, give rise to strong integrality, both for the collection of cuts induced by the crossing family and for the (blocking) collection of covers of the crossing family. This setting subsumes many of the examples of strong integrality discussed in [Schrijver 1982].

We now consider strong integrality results for discrete covering models. For covering models it is plain that when matrix A has integral entries, the strong min-max condition can hold only if A is a (0,1)-valued matrix. Thus we restrict attention to models for which A is (0,1)-valued and observe the well-known result that strong integrality holds here precisely when the (set-wise) maximal rows of A correspond to the maximal cliques in a perfect graph. The subject of perfect graphs is covered thoroughly by the following two recent references:

M.C. Golumbic (1980). *Algorithmic Graph Theory and Perfect Graphs,* Academic Press, New York;

C. Berge, V. Chvátal (eds.) (1984). *Topics on Perfect Graphs, Ann. Discrete Math. 21.*

Essentially two approaches have emerged for establishing strong integrality results such as those outlined above. The first we consider is algebraic in nature and is based on the concept of *total dual integrality,* first stated in full generality in

J. Edmonds, F.R. Giles (1977). A min-max relation for submodular functions on graphs. *Ann. Discrete Math. 1,* 185-204.

For packing models total dual integrality of the linear system $\{Ax \geqslant 1, x \geqslant 0\}$ arising from the linear programming dual of $P(A,c)$ is a restatement of the strong max-min stipulation relating $P(A,c)$ and $P_I(A,c)$, and similarly for covering models and systems of the form $\{Ax \leqslant 1, x \geqslant 0\}$. The use of total dual integrality as a tool for establishing combinatorial max-min and min-max statements (and hence for establishing strong integrality properties) is surveyed extensively in [Schrijver 1984] (see §1). Additional important references on the topic of total dual integrality are

F.R. Giles, W.R. Pulleyblank (1979). Total dual integrality and integer polyhedra. *Linear Algebra Appl. 25,* 191-196,

where it is shown that any integral polyhedron can be represented by a totally dual integral system of the form $\{Ax \leqslant b\}$ with b integral, and

A. Schrijver (1981). On total dual integrality. *Linear Algebra Appl. 38,* 27-32,

which establishes existence of a unique minimal totally dual integral system $\{Ax \leqslant b\}$ with A integral for representing a full dimensional rational polyhedron. In the latter case $\{x : Ax \leqslant b\}$ is integral if and only if b is integral.

One interesting and important consequence of total dual integrality of the system $\{Ax \geqslant 1, x \geqslant 0\}$ is that $\{x : Ax \geqslant 1, x \geqslant 0\}$ is an integral polyhedron (a similar statement holds for covering models). This was observed in [Fulkerson 1971] (see §1); generalizations of this result are proved in

A.J. Hoffman (1974). A generalization of max flow-min cut. *Math. Programming 6,* 352-359

and in [Edmonds & Giles 1977] (see above).

A second approach for establishing strong integrality results is algorithmic. In many of the examples cited above, a polynomial-time algorithm is known for solving $P_I(A,c)$ or $C_I(A,c)$ which yields strong integrality as a by-product. For such algorithms the reader can refer to, e.g.,

E.L. Lawler (1976). *Combinatorial Optimization: Networks and Matriods,* Holt, Rinehart and Winston, New York (for the max flow problem),

L. Lovász (1976). On two minimax theorems in graph theory. *J. Combin. Theory Ser. B 21,* 96-103 (for rooted spanning branchings and directed cutsets),

and to [Fulkerson 1971] (see §1), [Fulkerson 1974], [Frank 1979] and [Schrijver 1982] (see above) for algorithmic discussions concerning, respectively, (s,t)-cut positive parts, rooted cut positive parts, kernel systems and strong connectors. Finally, a major contribution of [Grötschel, Lovász & Schrijver 1981] (see §1) is the use of the ellipsoid algorithm to construct a polynomial-time algorithm for solving $C_I(A,c)$ when A is the (vertex-)incidence matrix of maximal cliques of a perfect graph.

It is important to point out that all the algorithms mentioned in the previous paragraph are polynomial-time in the input length required to describe the associated graph as opposed to the length required to describe the matrix A. The point here is that, for example in the case of a perfect graph G on n vertices, even though G may have exponentially (in n) many maximal cliques (rows of A), the algorithm from [Grötschel, Lovász & Schrijver 1981] for solving $C_I(A,c)$ runs in time which is a polynomial function of n and the length required to describe the vector c. If we do consider the matrix A as the given data, however, then it has been shown in

S. Baum, L.E. Trotter, Jr. (1982). Finite checkability for integer rounding properties in combinatorial programming problems. *Math. Programming 22,* 141-

147

that the optimal values of $P(A,c)$ and $P_I(A,c)$ are equal for all nonnegative integral vectors c if and only if equality holds for a certain easily described *finite* set of nonnegative integral c, and similarly for covering. Hence strong integrality for a given matrix A can be verified in finite time. More generally, building on the algorithmic approach in

R. Chandrasekaran (1981). Polynomial algorithms for totally dual integral systems and extensions. *Ann. Discrete Math. 11,* 39-51,

Cook has shown that recognition of whether a given linear system is totally dual integral is a problem in co-$\mathcal{NP}$. This result appears in

W. Cook (1982). *Recognition of Totally Dual Integral Systems,* CORR report 82-20, University of Waterloo, Ontario.

These computational complexity results extend naturally to the setting of integer rounding, which is the topic of the following section.

3. Integer Rounding

Suppose we are given a nonnegative matrix A and a nonnegative vector c for which the optimum value of $P(A,c)$ is not an integer. Then strong integrality fails for this discrete packing model, but it is reasonale to ask whether the optimum values of $P(A,c)$ and $P_I(A,c)$ remain 'close'. Thus it is said that a discrete packing model has the *integer round down property* when, for any nonnegative integral vector c, the optimum value of $P_I(A,c)$ is given by the largest integer which does not exceed the value of $P(A,c)$. An *integer round up property* is defined analogously for discrete covering models. In this section we indicate several packing and covering models which have these properties; we mention again that a survey of such models appears in [Marcotte 1983] (see §1).

Perhaps the best-known integer rounding results arise when A is the incidence matrix of bases in a matroid. Then integer rounding holds for both packing and covering by the rows of A; this is a consequence of the work in

J. Edmonds (1965). Minimum partition of a matroid into independent subsets. *J. Res. Nat. Bur. Standards 69B,* 67-72,

J. Edmonds, D.R. Fulkerson (1965). Transversals and matroid partition. *J. Res. Nat. Bur. Standards 69B,* 147-153.

These results are extended in

S. Baum, L.E. Trotter, Jr. (1981). Integer rounding for polymatroid and branching optimization problems. *SIAM J. Algebraic Discrete Meth. 2,* 416-425

to the setting in which the rows of A correspond to the bases of an integral polymatroid. The approach in this paper is algebraic, using (local) total unimodularity to establish a form of polyhedral integral decomposition (see below), whereas in [Edmonds 1965] (see above) a polynomial-time algorithm is given for covering the elements of a matroid by its bases. In [Marcotte 1983] (see §1) a min-max result for machine scheduling presented in

T.C. Hu (1961). Parallel sequencing and assembly line problems. *Oper. Res. 9,* 841-848

is derived from the integer round up property for matroid basis covering.

Next suppose we are given an integral supply-demand network (all supply, demand and capacity data are integral) and that the rows of A are the integral feasible (edge-)flows of this network. It is shown in

D.R. Fulkerson, D.B. Weinberger (1975). Blocking pairs of polyhedra arising from network flows. *J. Combin. Theory Ser. B 18,* 265-283

that integer round down holds for the associated discrete packing model. Note that the special case of one source, one sink with unit supply, demand and capacities corresponds to the maximum flow model considered in the previous section. Similiar results are obtained in the same paper for uncapacitated integral supply-demand networks (using minimal integral feasible flows), and in

D.B. Weinberger (1976). Network flows, minimum coverings, and the four-color conjecture. *Oper. Res. 24,* 272-290

corresponding integer round up results are developed for the analogous covering models. These packing and covering results are extended in

L.E. Trotter, Jr., D.B. Weinberger (1978). Symmetric blocking and anti-blocking relations for generalized circulations. *Math. Programming Stud. 8,* 141-158

to models defined by matrices whose rows consist of the integral solutions to linear systems of the form $\{Nx = 0, a \leqslant x \leqslant b\}$, where N is a totally unimodular matrix and $0 \leqslant a \leqslant b$ with vectors a and b integral. The results of these three references are established algebraically; in [Marcotte 1983] (see §1) and in

R.E. Bixby, O.M.-C. Marcotte, L.E. Trotter, Jr. (to appear). *Packing and Covering with Integral Feasible Flows of Integral Supply-Demand Networks,*

polynomial-time (in the size of the network data) algorithms are given which can be used to solve $P_I(A,c)$ and $C_I(A,c)$ in the network cases.

In special cases the incidence matrix of certain common independent sets for two matroids (defined on the same ground set) exhibits integer rounding properties. When the rows of A correspond to the maximum cardinality common independent sets of two strongly base orderable matroids, integer rounding results for packing and covering are obtained in

C.J.H. McDiarmid (1976). *On Pairs of Strongly-Base-Orderable Matroids,* Technical report 283, School of Operations Research and Industrial Engineering, Cornell University, Ithaca, NY.

Integer round up results are also obtained in this paper for the case in which the rows of A give the incidence of (set-wise) maximal common independent sets of two strongly base orderable matroids; for general matroids these results fail. In

M.D. McDaniel (1981). *Network Models for Linear Programming Problems with Integer Rounding Properties,* M.S. thesis, School of Operations Research and Industrial Engineering, Cornell University, Ithaca, NY,

it is shown that similar results for two gammoids (a class of matroids properly subsumed by strongly base orderable matroids) can be derived from the model of [Fulkerson & Weinberger 1975] (see above) by consideration of an appropriate supply-demand network, thus tracing these integrality results back, in an algebraic sense, to total unimodularity of the vertex-edge incidence matrix of a digraph. The approach of [McDiarmid 1976] is algorithmic, relying on earlier work in

J. Davies, C.J.H. McDiarmid (1976). Disjoint common transversals and exchange structures. *J. London Math. Soc. (2) 14,* 55-62.

Branchings provide another 'matroid intersection' example for which integer rounding properties hold. Integer round down for the family of maximum cardinality branchings in a digraph and integer round up for both this family and the family of maximal branchings are established in [Baum & Trotter 1981] (see above) using Edmonds' 'edge-disjoint (rooted) branchings theorem' (see [Edmonds 1972] in §2). Integer round up for the case of rooted spanning branchings (the covering analogue of Edmonds' packing result in [Edmonds 1972]) follows from min-max results in [Frank 1979] (see §2) and in

K. Vidyasankar (1978). Covering the edge set of a directed graph with trees. *Discrete Math. 24,* 79-85,

A. Frank (1979). Covering branchings. *Acta Sci. Math. (Szeged) 41,* 77-81.

This can also be deduced from the results in [Baum & Trotter 1981] (see above).

As a final example we mention that in [Marcotte 1983] above the integer round up property is shown to hold for certain classes of cutting stock problems. Note that the usual formulation of the cutting stock problem is as a discrete covering model for which the rows of matrix A are the integral feasible solutions to a knapsack problem.

The integer rounding properties for packing and covering models are equivalent to a type of integral decomposition of related polyhedra (see [Baum & Trotter 1981] above). Thus integral decomposition provides a useful means for establishing integer rounding results. Several variations of the notion of integral decomposition, as well as an indication of certain combinatorial models for which these alternative refinements hold, are presented in

C.J.H. McDiarmid (1983). Integral decomposition in polyhedra. *Math. Programming 25,* 183-198.

We mention again that the computational complexity results of [Baum & Trotter 1982], [Chandrasekaran 1981] and [Cook 1982] (see §2) remain valid in the integer rounding framework and we add that in

J. Orlin (1982). A polynomial algorithm for integer programming covering problems satisfying the integer round-up property. *Math. Programming 22,* 231-235,

co-$\mathcal{NP}$ recognition of the integer rounding properties for discrete packing and covering models was first established.

4. An Open Question

For certain combinatorial families of interest a slight weakening of the notion of integer rounding may hold. In this section we indicate such a possibility for the edge-coloring problem on an undirected graph. Suppose A is the (edge-) incidence matrix of matchings in a simple undirected graph G. Then $C_I(A,1)$ is the problem of determining a minimum coloring of the edges of G. One can verify that when G is, for example, the Petersen graph (see

J.A. Bondy, U.S.R. Murty (1976). *Graph Theory with Applications,* North-Holland, New York),

the values of $C(A,1)$ and $C_I(A,1)$ differ by unity. Thus integer round up does *not* hold for this discrete covering problem. Nevertheless, Vizing's Theorem (see [Bondy & Murty 1976]) states that for any simple graph G the minimum number of colors required to color the edges of G exceeds the maximum degree of a vertex in G by at most 1, which implies that the values of $C_I(A,1)$ and $C(A,1)$ differ by *at most unity.* The latter assertion follows because, for any nonnegative integral vector c,

$$\min\{1\cdot y : yA \geqslant c, y \geqslant 0, y \text{ integral}\}$$

$$\geqslant \tag{1}$$

$$\min\{1\cdot y : yA \geqslant c, y \geqslant 0\}$$

$$= \tag{2}$$

$$\max\{c\cdot x : Ax \leqslant 1, x \geqslant 0\}$$

$$\geqslant \tag{3}$$

$$\max\{c\cdot x : x \text{ is the incidence vector of a star in G}\},$$

where relation (1) is obvious, relation (2) follows from linear programming duality theory and (3) is valid because any incidence vector of the star of a vertex in G satisfies the linear system $\{Ax \leqslant 1, x \geqslant 0\}$. Vizing's Theorem for simple graphs thus insures that all the above expressions differ by at most unity when $c = 1$.

To what extent is the preceding development valid for multigraphs, i.e., for arbitrary nonnegative integral c in the above expressions? Vizing's Theorem for a multigraph G (see [Bondy & Murty 1976]) asserts that the difference between the size of a minimum edge coloring and that of the largest star in G is at most the largest multiplicity of an edge in G; it is easy to construct examples for which this maximum possible difference is achieved. Thus for general c the first and last expressions above may differ by as much as the largest component of c. In

P.D. Seymour (1979). On multi-colourings of cubic graphs, and conjectures of Fulkerson and Tutte. *Proc. London Math. Soc. 38,* 423-460,

Seymour raises the question of whether for general c the difference between $C_I(A,c)$ and $C(A,c)$, i.e., the difference governing relation (1) above, remains *at most unity.* Resolving this question seems to be quite difficult, but were it to be settled in the affirmative, edge coloring would provide a combinatorial model for which a natural weakening of the integer round up property would hold. We conclude by recalling that an integer rounding result is often accompanied by a polynomial-time algorithm for solving the associated discrete packing or covering problem. In the present instance, however, Holyer has shown in

I. Holyer (1981). The *NP*-completeness of edge-colouring. *SIAM J. Comput. 10,* 718-720

that the edge coloring problem is $\mathcal{NP}$-complete.

4

Submodular Functions and Polymatroid Optimization

E.L. Lawler
University of California, Berkeley

CONTENTS

A set function $f : 2^E \rightarrow \mathbb{R}$ is said to be *submodular* if, for all subsets X, Y of E,

$$f(X)+f(Y) \geqslant f(X \cup Y) + f(X \cap Y).$$

If a discrete optimization problem can be solved efficiently, it is quite likely that submodularity is responsible. In recent years there has been a growing appreciation of the fact that submodularity plays a pivotal role in discrete optimization, not unlike that of convexity in continuous optimization.

Submodular functions are perhaps best known for their role in matroid theory. A *matriod* is defined by an integer-valued submodular function with these additional properties:

$$f(\varnothing) = 0;$$

$$f(\{e\}) = 0 \text{ or } 1, \quad \text{for all } e \in E;$$

$$f(X) \leqslant f(Y), \quad \text{for all } X \subset Y \subset E.$$

A *polymatroid* is obtained by dropping the requirement that $f(\{e\}) = 0$ or 1.

This bibliography is intended to give the reader a number of references that will provide an introduction to the subject of submodular functions, polymatroids, and the optimization problems that are associated with them.

1. BACKGROUND MATERIAL, SURVEYS, ETC.

E.L. Lawler (1976). *Combinatorial Optimization: Networks and Matroids,* Holt Rinehart and Winston, New York.

This book introduces the theory of matroids from the viewpoint of their role in combinatorial optimization. It provides a number of references to the literature, particularly to the pioneering works of Jack Edmonds, through the time of publication. Among the algorithms described are primal and primal-dual algorithms for the weighted matroid intersection problem. The present bibliography concentrates on publications that have appeared since the time of this book.

D.J.A. Welsh (1976). *Matroid Theory,* Academic Press, London.

The standard text on matroid theory, from the viewpoint of pure mathematics. As in the case of Lawler's book, some of the material has been outdated by more recent developments.

R. Bixby (1981). Matroids and operations research. H.J. Greenberg, F.H. Murphy, S.H. Shaw (eds.). *Advanced Techniques in the Practice of Operations Research,* North-Holland, New York, 333-458.

This survey paper gives an excellent introduction to various aspects of matroid theory that are not well covered in Lawler's book. Of particular interest is Bixby's discussion of linear representation of matroids and the fields over which matroids are representable. This paper also gives an excellent introduction to P. Seymour's impressive polynomial algorithm for testing matrices for total unimodularity.

L. Lovász (1983). Submodular functions and convexity. A. Bachem, M. Grötschel, B. Korte (eds.). *Mathematical Programming: the State of the Art - Bonn 1982,* Springer, Berlin, 235-257.

This stimulating and very readable survey explains why submodular functions are like convex functions (except when they are more like concave functions). Lovász also discusses some of the more useful operations on submodular functions, describes the polyhedra defined by submodular functions and formulates mathematical programming problems with submodular objective functions and constraints.

A. Schrijver (1984). Total dual integrality from directed graphs, crossing families, and sub- and supermodular functions. W.R. Pulleyblank (ed.). *Progress in Combinatorial Optimization,* Academic Press, New York, 315-361.

The theory of total dual integrality is one of the more significant recent developments in combinatorial optimization. This paper provides a good introduction to the relation between TDI systems and submodular functions.

2. Polymatroidal Network Flows

E.L. Lawler, C.U. Martel (1982). Computing 'maximal' polymatroidal network flows. *Math. Oper. Res. 7,* 334-347.

It is shown how many (if not all, except the matroid parity problem) of the optimization problems involving polymatroids and submodular functions can be formulated and solved by generalizing the classical network flow model to permit submodular capacity constraints to be applied to sets of arcs. The augmenting path theorem, the integrality theorem and the max-flow min-cut theorem of classical network flow theory all have immediate generalizations in this model. Moreover, the authors are able to show that there is an analog to the Edmonds-Karp bound on the number of flow augmentations that are required to obtain a maximum value flow, provided that at each step one chooses an augmenting path that is not simply shortest in number of arcs but is also shortest with respect to a simple rule of lexicography. (The significance of the rule of lexicography was discovered independently by P. Schonsleben in his doctoral thesis, ETH Zürich, 1980.)

E.L. Lawler, C.U. Martel (1982). Flow network formulations of polymatroid optimization problems. *Ann. Discrete Math. 16,* 189-200.

Just as the König-Egerváry theorem of bipartite matching can be obtained as a simple corollary of the max-flow min-cut theorem of network flows, so too can many duality results be obtained as simple corollaries of the max-flow min-cut theorem of polymatroidal network flows. Among these duality results are the polymatroidal intersection duality theorem, the Rado-Hall theorem, and duality theorems of linking systems.

E.L. Lawler (1982). *Generalizations of the Polymatroidal Network Flow Model,* Report BW 158, Centre for Mathematics and Computer Science, Amsterdam.

It is shown how the polymatroidal flow model can be generalized in various ways: to accommodate so-called submodular intersecting and crossing families of capacity constraints, to provide for lower bounds on arc flow determined by supermodular set functions. (A function is *supermodular* if its negative is submodular.) It is shown how to compute a feasible flow in a network with both submodular capacities and supermodular lower bounds, provided the submodular functions and supermodular functions are in a relation of *compliance*. As a byproduct of this development, a particularly simple and elegant proof of the *discrete separation* theorem of Andras Frank is given. Finally, it is shown how the Edmonds-Giles model of submodular optimization can be formulated and solved as a polymatroidal network flow problem. (More about the Edmonds-Giles model below.)

C.A. Tovey, M.A. Trick (1983). *An $O(m^4 d)$ Algorithm for the Maximum Polymatroidal Flow Problem,* School of Industrial and Systems Engineering, Georgia Institute of Technology, Atlanta.

The authors improve on the maximum flow computation of Lawler and Martel by using an approach that is analogous to Dinic's layered network improvement of the Edmonds-Karp computation for the classical flow model. They succeed in obtaining an improvement by a factor of n (where n is the number of nodes) in the worst-case running time of the computation. It seems unlikely, however, that the polymatroidal flow computation admits of further improvement along the lines of Karzanov.

R. Hassin (1982). Minimum cost flow with set constraints. *Networks 12,* 1-21.

The polymatroidal flow model was developed independently by Raphael Hassin, who has provided algorithms for minimum cost flow versions of the model. It appears that there is an analog for polymatroidal flows of each of the min-cost flow algorithms that are known for the classical flow model.

3. Equivalent Computational Models

J. Edmonds, R. Giles (1977). A min-max relation on submodular functions on graphs. *Ann. Discrete Math. 1,* 185-204.

Besides the polymatroidal network flow model of Hassin and Lawler and Martel, there are many other models which enable one to formulate essentially the same class of optimization problems. Edmonds and Giles suggest a model in which conservation of flow is not required at each node. Instead, the net flow into a set of nodes is constrained by a submodular function. On the other hand, capacities are applied only to individual arcs, rather than to sets of arcs, as in the polymatroidal network flow model. It is possible to argue which of these models is 'better'; preference is largely a matter of esthetics and, possibly, which model one becomes familiar with first.

S. Fujishige (1978). Algorithms for solving independent flow problems. *J. Oper. Res. Soc. Japan 21,* 189-204.

Another alternative model: the net flow out of *source* nodes and the net flow into *sink* nodes are constrained by submodular functions on those sets of nodes. It is a simple exercise to transform this model into the polymatroidal flow model.

4. Oracles and Polymatroid Feasibility

W.H. Cunningham (1984). Testing membership in matroid polyhedra. *J. Combin. Theory Ser. B. 36,* 161-188.

It is a generally accepted convention to describe polymatroid optimization algorithms in terms of *oracles* or subroutines that answer certain crucial questions in the course of the computation. For example, in the case of the polymatroidal network flow model it is necessary to repeatedly determine whether or not the flow through a given set of arcs is feasible with respect to the capacity constraints. Just how such a question should be answered depends very much on the underlying structure of the submodular constraints and how they are presented for computational purposes.

Cunningham provides a satisfying solution to the problem, for the special case that the capacity constraints are matroidal in character. The general problem that remains open is equivalent to the following: find the minimum of an arbitrary submodular function with a number of evaluations of the function that is bounded by a polynomial in the number of elements in the ground set over which the function is defined. (Grötschel, Lovász and Schrijver have shown that there is a polynomial bounded algorithm, based on the ellipsoid method of linear programming. What is wanted here is a *direct* combinatorial algorithm.)

D. Hausmann, B. Korte (1980). The relative strength of oracles for independence systems. J. Frehse *et al.* (eds.). *Special Topics of Applied Mathematics,* North-Holland, Amsterdam, 195-211.
G.C. Robinson, D.J.A. Welsh (1980). The computational complexity of matroid properties. *Math. Proc. Cambridge Philos. Soc. 87,* 29-45.

The above two papers deal with the relative strength of computational oracles. Various matroidal properties are shown to be polynomially equivalent, in the sense that a polynomial number of calls on an oracle for one property can be used to answer questions for the other. Some oracles are shown to be strictly stronger than others. Issues concerning oracles can sometimes be very significant, as in the case of the matroid parity problem (see below).

5. The Matroid Parity Problem

L. Lovász (1980). Matroid matching and some applications. *J. Combin. Theory Ser. B 28,* 121-131.
L. Lovász (1981). The matroid matching problem, L. Lovász, V. T. Sós (eds.). *Algebraic Methods in Graph Theory,* North-Holland, Amsterdam, 495-517.

In these very significant papers Lovász resolves the *matroid parity* (or *matroid matching*) problem, the status of which had been an open question for a fairly long time. (See Ch. 9 of Lawler's book.) He first shows that the general case of the problem cannot be solved with a polynomial number of calls on an appropriate oracle. He then shows that the problem can be solved in polynomial time, provided the matroid has a linear representation. An easy formulation of this special case is as follows: given a set of n pairs of vectors over an arbitrary field, find a maximum number of pairs such that their union is a linearly independent set. A still more special case is as follows: given an arbitrary undirected graph whose edges are paired, find a maximum number of pairs, such that their union does not contain the edges of any cycle. That is, the edges chosen yields a forest of trees. Lovász's algorithm is very complex in all senses of the word - comprehension, coding, and worst-case number of operations. In spite of this, or perhaps because of this, the algorithm is a major achievement.

J. Orlin, E. Gugenheim, J. Hammond, J. VandeVate (1983). *Linear Matroid Parity Made Almost Easy: an Extended Abstract,* Sloan School of Management, Massachusetts Institute of Technology, Cambridge.

The authors propose an algorithm that is much simpler and more elegant than that of Lovász. Moreover, it has the virtue of indicating a close relation between the matroid parity problem and graphic matching, one of the problems that it clearly generalizes. Unfortunately, this paper gives only a fragmentary sketch of the proposed computational procedures. The complete algorithm will be described in the forthcoming doctoral thesis of VandeVate.

M. Stallman (1983). *An Augmenting Paths Algorithm for the Matroid Parity Problem on Binary Matroids,* Ph.D. thesis, University of Colorado, Boulder.

Stallman considers the case of the parity problem in which the matroid is *binary*, that is, representable over the field of two elements. He shows that there is a computational procedure in which sets are augmented by *augmenting sequences* of elements (cf. Ch. 9 of Lawler's book) much like augmenting paths in network flow theory.

Po Tong, E.L. Lawler, V.V. Vazirani (1984). Solving the weighted parity problem for gammoids by reduction to graphic matching. W.R. Pulleyblank (ed.). *Progress in Combinatorial Optimization,* Academic Press, New York, 363-374.

Lovász's algorithm solves only the unweighted case of the matroid parity problem; the weighted case for linearly representable matroids remains an open problem at the present time. In this paper it is shown how the weighted parity problem can be solved for *gammoids* by a very simple reduction to weighted graphic matching. The matroid of a *series parallel* network is a gammoid. Hence the weighted paired-edges problem (see above) for that class of graphs can be solved by graphic matching techniques. At the present time, it is not known how to solve any other weighted parity problems in polynomial time, except weighted graphic matching itself.

6. Applications

M. Iri (1982). Applications of matroid theory. A. Bachem, M. Grötschel, B. Korte (eds.). *Mathematical Programming: the State of the Art - Bonn 1982,* Springer, Berlin, 158-201.

Various applications of optimization involving submodular functions are reviewed, with emphasis on the electrical network problems. Attention is given to the problem of *principal partition* of matroids, an approach much favored by the Japanese.

L. Lovász, Y. Yemini (1982). On generic rigidity in the plane. *SIAM J. Algebraic Discrete Meth. 3,* 91-99.

Questions of structural rigidity are among the most interesting that can be dealt with by polymatroid optimization. This paper provides an introduction to this subject.

C.U. Martel (1982). Preemptive scheduling with release times, deadlines, and due times. *J. Assoc. Comput. Mach. 29,* 812-829.

Polymatroidal network flows provide a particularly elegant and effective means for formulating and solving the preemptive parallel machine scheduling problem dealt with in this paper. As a matter of fact, it was this problem that suggested the polymatroidal flow formulation to Lawler and Martel.

7. Generalizations of Matroids and Polymatroids

B. Korte, L. Lovász (1981). Mathematical structures underlying greedy algorithms. F. Gécseg (ed.). *Fundamentals of Computation Theory,* Lecture Notes in Computer Science 117, Springer, Berlin, 205-209.

It is well known that the *greedy algorithm* works for weighted matroids and, in a sense, only for matroids. Yet there are structures (*greedoids*) for which *greedy-like* algorithms work. This paper axiomatizes one such class of structures and indicates some of their properties.

A. Frank (to appear). Generalized polymatroids, *Proc. 6th Hungarian Combinatorial Colloquim, Eger, 1981.*
E. Tardos (1983). *Generalized Matroids and Supermodular Colourings,* Report AE 19/83, Faculty of Actuarial Sciences and Econometrics, University of Amsterdam.

Frank and Tardos suggest another generalization of matroids for which the greedy algorithm and various duality results of matroid theory are valid.

5

Computational Complexity

C.H. Papadimitriou
Stanford University &
National Technical University of Athens

CONTENTS

1. INTRODUCTION

The last decade has seen a true explosion of activity, ideas, and results in the field of computational complexity, the area of the theory of computation that studies the inherent limitations of the performance of algorithms. A lot of the excitement (and a great part of the results) comes from the applications of this theory in combinatorial optimization, and we list many references below in witness of this. There are in fact several introductions to computational complexity that were written with this community at least partly in mind; see

R.M. Karp (1975). On the complexity of combinatorial problems. *Networks 5*, 44-68,

J.D. Ullman (1975). Complexity of sequencing problems. E.G. Coffman, Jr. (ed.). *Computers & Job/Shop Scheduling Theory,* Wiley, New York,

M.R. Garey, D.S. Johnson (1979). *Computers and Intractability: a Guide to the Theory of NP-Completeness,* Freeman, San Fransisco,

C.H. Papadimitriou, K. Steiglitz (1982). *Combinatorial Optimization: Algorithms and Complexity,* Prentice-Hall, Englewood Cliffs, NJ,

D.S. Johnson, C.H. Papadimitriou (1985). Computational complexity. E.L. Lawler, J.K. Lenstra, A.H.G. Rinnooy Kan, D.B. Shmoys (eds.). *The Traveling Salesman Problem,* Wiley, Chichester.

These introductions, however, fail — because of pedagogical considerations and constraints of space and scope — to give a complete picture of the underlying issues and related previous work in computational complexity. In the interest of helping the two research communities approach each other and develop a common language, it seems appropriate to bring to the attention of workers in combinatorial optimization a more complete view of the prehistory, history, and state of research in computational complexity. The present annotated bibliography is an attempt in this direction. Anybody who has even a passing acquaintance with the field realizes that such a bibliography cannot be anywhere close to complete. Our approach has been to emphasize as many aspects of computational complexity as possible, perhaps being incomplete in our coverage of each aspect. If in some area there is already a review in the literature, or a recent work refers comprehensively to the previous literature, we have omitted original references. Wherever good and complete reviews already exist in the literature (examples of such areas are the theory and methodology of $\mathcal{NP}$-completeness, because of [Garey & Johnson 1979], and the complexity of parallel computation, with the bibliography that appears elsewhere in this volume), we have been particularly light in our selection of references. Our only claim to completeness (and an easy one to make) is that the vast majority of the literature on computational complexity lies at depth two, referenced by our references.

2. The Origins

Establishing the intellectual family tree of a field is always a nearly impossible task. It seems rather uncontroversial, however, that the origins of computational complexity can be traced back to the brilliant first attempt at a theory of computation, namely the *theory of computability*, which was initiated by A.M. Turing in

A.M. Turing (1937). On computable numbers, with an application to the Entscheidungsproblem. *Proc. London Math. Soc. (2) 42,* 230-265.

Good treatments of this theory, which studied whether problems can have an algorithm *at all*, efficient or not, can be found in

S.C. Kleene (1952). *Introduction to Metamathematics,* Van Nostrand, New York,

H. Rogers, Jr. (1967). *The Theory of Recursive Functions and Effective Computability,* McGraw Hill, New York,

M. Machtey, P.R. Young (1978). *An Introduction to the General Theory of Algorithms,* Elsevier, New York,

H.R. Lewis, C.H. Papadimitriou (1981A). *Elements of the Theory of Computation,* Prentice-Hall, Englewood Cliffs, NJ.

The latter two textbooks also treat the relation between computability and complexity. It is interesting to note that even in early works on computability one already finds references to the need of more refined measures of the 'effectiveness' of algorithms.

Another related source of influence has been the *theory of formal languages,* developed in connection to efforts to understand natural and computer languages. Formal languages – that is, infinite sets of strings over some fixed alphabet – were categorized according to their *structural* (not yet computational) complexity, that is, the complexity of the description of their generators and acceptors. The classical work here is that of Chomsky in the 1950's, e.g.

N. Chomsky (1956). Three models for the description of language. *IRE Trans. Inform. Theory 2,* 113-124,

N. Chomsky (1959). On certain formal properties of grammars. *Inform. and Control 2,* 137-167.

The concept of *completeness*, for example, so useful in complexity these days, has its origins in both recursion and language theory; see [Rogers 1967] above and, e.g.,

S.A. Greibach (1973). The hardest context-free language. *SIAM J. Comput. 2,* 304-310.

A parallel *theory of automata* was developed, from which complexity theory was to borrow heavily. See

M.O. Rabin, D. Scott (1959). Finite automata and their decision problems. *IBM J. Res. Develop. 3,* 114-125.

Some books on these topics:

M.A. Harrison (1978). *Introduction to Formal Language Theory,* Addison-Wesley, Reading, MA,

J.E. Hopcroft, J.D. Ullman (1979). *Introduction to Automata Theory, Languages and Computation,* Addison-Wesley, Reading, MA.

The latter book treats a wider range of topics, including computational complexity.

On the other hand, the period of fermentation that forced computational complexity into existence, started with the development of real digital computers (as opposed to the hypothetical models used by the mathematicians in the 1930's), and the many applications that started to become feasible. Scientists using the computers started wondering about more scientific ways of analyzing and predicting their performance. The most direct influence came from two fields that were in a sense 'born' with the computer: *numerical analysis* and *operations research.* For an account of algorithmic and complexity issues related to numerical computation, see

D.E. Knuth (1981). *The Art of Computer Programming; Volume 2: Semi-Numerical Algorithms; Second Edition,* Addison-Wesley, Reading, MA,

A. Borodin, I. Munro (1975). *The Complexity of Algebraic and Numeric Problems,* Elsevier, New York.

Remarkably, the first conscious reference to the notion of a polynomial-time algorithm (which is currently dominating the field of computational complexity) came from an intellectual giant who did some of the most original work in both computer science and optimization:

J. von Neumann (1953). A certain zero-sum two person game equivalent to the optimal assignment problem. H.W. Kuhn, A.W. Tucker (eds.). *Contributions to the Theory of Games II,* Princeton University Press, Princeton, NJ.

For other examples of early complexity-related worries in optimization see

G.B. Dantzig, D.R. Fulkerson, S.M. Johnson (1954). Solution of a large scale travelling salesman problem. *Oper. Res. 2,* 393-410,

M. Held, R.M. Karp (1962). A dynamic programming approach to sequencing problems. *J. SIAM 10,* 196-210,

G.B. Dantzig (1963). *Linear Programming and Extensions,* Princeton University Press, Princeton, NJ,

J. Edmonds (1965). Paths, trees, and flowers. *Canad. J. Math. 17,* 449-467.

Several central concepts of today's complexity theory, such as polynomial time as well as $\mathcal{NP}$, are first hinted at in the work of Edmonds during this period.

In the meantime, researchers in the theory of computation and logic were closing in towards complexity. There were examples of structural restrictions on languages (in the spirit of language or recursion theory) that turned out to be complexity classes (albeit too powerful to be of practical interest). Some examples are the hierarchies of Grzegorczyk and Axt; see

R.W. Ritchie (1963). Classes of predictably computable functions. *Trans. Amer. Math. Soc. 106,* 139-173,

A.R. Meyer, D.M. Ritchie (1967). The complexity of loop programs. *Proc. 22nd National ACM Conf.,* 465-469,

A. Cobham (1964). The intrinsic computational difficulty of functions. *Proc. 1964 Internat. Congress Logic, Methodology and Philosophy of Science,* North-Holland, Amsterdam, 24-30.

In fact, the class $\mathcal{P}$ (of polynomial-time solvable problems) was first defined formally in the last paper. However, the era of computational complexity starts with these pioneering papers, that stated explicitly the issues and the techniques to attack them:

M.O. Rabin (1960). *The Degree of Difficulty in Computing a Function and a Partial Order of Recursive Sets,* Report 2, Branch of Applied Logic, Hebrew University, Jerusalem,

J. Hartmanis, R.E. Stearns (1965). On the computational complexity of algorithms. *Trans. Amer. Math. Soc. 117,* 285-305.

This was a critical moment for complexity theory, and seemingly insignificant details of the mathematical formalism had a deep influence on its development. For example, the latter paper contained the seeds of the now prevailing preference for asymptotic analysis and the liberal treatment of constants, in that the authors assumed their machines to have a finite but unbounded number of symbols.

3. Machines, Measures, and Modes

In computational complexity one starts with a computational *model* (a machine capable of carrying out algorithms) and a *resource* (something expensive the machine uses up). The most usual resource is *time,* measured in the number of elementary steps the machine takes to carry out the computation on a given input. This number is maximized over inputs of the same length, and the resulting function of the length of the input is the time complexity of the computation. Another useful resource is *space*, that is, the amount of internal storage the machine must use for the computation (ignoring the space taken up by the input). Other kinds of resources, however, have been studied.

There are many models of computation that have been proposed in the literature as the basis of computational complexity. The original machine on which complexity issues were first discussed was the *Turing machine* ([Hartmanis & Stearns 1965], see §2), a rather primitive and clumsy computer, which has a simple structure facilitating mathematical reasoning, and is nevertheless as powerful as any computer (see, e.g., [Lewis & Papadimitriou 1981A] in §2 for an introduction to these machines). There are several variants of the Turing machine. A Turing machine can have one or several tapes (and the tapes may have dimension higher than one!). Also, there is a model that is closer to the von Neumann computer, called the *random access machine* (RAM), introduced and studied in

J.C. Shepherdson, H.E. Sturgis (1963). Computability of recursive functions. *J. Assoc. Comput. Mach. 10,* 217-255,

S.A. Cook, R.A. Reckhow (1973). Time-bounded random access machines. *J. Comput. System Sci. 7,* 354-375.

See also

A.V. Aho, J.E. Hopcroft, J.D. Ullman (1974). *The Design and Analysis of Computer Algorithms,* Ch. 1, Addison-Wesley, Reading, MA,

A. Bertoni, G. Mauri, N. Sabadini (1981). A characterization of the class of functions computable in polynomial time by random access machines. *Proc. 13th Annual ACM Symp. Theory of Computing,* 168-176.

There is a real wealth of results on how fast these models can simulate one another. For the classical ones see, e.g., Chapter 12 of [Hopcroft & Ullman 1979] (§2), or Chapter 7 of [Lewis & Papadimitriou 1981A] (§2). These differences are not deemed terribly important, though, and in the sequel we shall adopt the Turing machine with many one-dimensional tapes as our 'standard' machine. The real reason why these different models are all 'the same' is that they can all simulate each other with only a *polynomial* waste of time - and polynomiality is what matters in complexity today...

This is not all, though. Besides machines and resources, we can vary our *mode* of computation. Our standard mode is the deterministic one - that of ordinary computers. There is another important mode, called *nondeterminism*. In this mode, the machine acquires the magical power to guess and always be lucky. Viewed otherwise, nondeterminism means that a machine can duplicate itself many times, and assign different tasks to its clones, each of which can be the final solver of the problem. This notion was apparently introduced in [Rabin & Scott 1959] (see §2), and was to be of major importance in complexity. A combination of a mode, resource, and bound gives us a complexity class. Examples: $\mathfrak{N}$SPACE(n) is the set of problems solvable by nondeterministic machines (the $\mathfrak{N}$ signifies nondeterminism) in space linear in the length of the input; TIME(2^n) is the class of problems solvable in deterministic (no $\mathfrak{N}$) exponential time. And so on. For an attempt to unify the various modes of computation, see

Hong Jia-wei (1981). On similarity and duality of computation. *Proc. 22nd Annual IEEE Symp. Foundations of Computer Science,* 348-359.

An immediate problem is to relate these classes by possible inclusions. There are many results in this respect. We shall henceforth assume that our bounds are 'constructible' functions, in that they are themselves computable in reasonable time by a machine (it turns out that this assumption is met by all 'ordinary' bounds that we may encounter in practice, but rules out some anomalies to be referred to later). Then, it can be shown that, if $F(n)$ is such a function growing faster than $G(n)$ (that is, when their quotient is unbounded), then TIME($G(n)$) is a proper subset of TIME($F(n)$). See

M. Fuerer (1982). The tight deterministic time hierarchy. *Proc. 14th Annual Symp. Theory of Computing,* 8-16.

This is a 'hierarchy' result, because it tells us that there is an infinite hierarchy of classes of problems, and by increasing our time bound by any meaningful amount, we shall get new problems that we could not solve before. Incidentally, one of the great frustrations of complexity theory is that, so far, only 'unnatural' or 'too powerful' problems have been shown to be in the difference of reasonably neighboring in this hierarchy time complexity classes. For some interesting examples of problems that have been shown to be intractable, see

A.R. Meyer, L.J. Stockmeyer (1973). The equivalence problem for regular expressions with squaring requires exponential space. *Proc. 13th Annual IEEE Symp. Switching and Automata Theory,* 125-129,

M.J. Fischer, M.O. Rabin (1974). Super-exponential complexity of Presburger arithmetic. R.M. Karp (ed.). *Complexity of Computation,* SIAM-AMS.

Hierarchy results also exist for space, and nondeterministic time and space; see

R.E. Stearns, J. Hartmanis, P.M. Lewis III (1965). Hierarchies of memory-limited computations. *Proc. IEEE Conf. Switching Theory,* 179-190,

J.I. Seiferas (1977). Techniques for separating space complexity classes. *J. Comput. System Sci. 14,* 73-129.

Our weakest point is relating classes of different kinds - comparing TIME(n^3) and $\mathcal{N}$SPACE(n), say. Naturally, there are some trivial results stating that, say, TIME(F) is a subset of SPACE(F), which in turn is a subset of TIME(c^F) (when constants like c here are mentioned in complexity classes, the class denoted is the union of all these classes, over all natural numbers c). Unfortunately, there are precious few nontrivial results here. The first one is due to Savitch, and states that, for space, we can get rid of unrealistic nondeterminism with only a quadratic waste. Thus, $\mathcal{N}$SPACE($F(n)$) is a subset of SPACE($F(n)^2$). See

W.J. Savitch (1970). Relationships between nondeterministic and deterministic tape complexities. *J. Comput. System Sci. 4,* 177-192.

This result differentiates space from time, because in the latter we think the waste has to be exponential. Also, another curious mode of computation, called *symmetry*, seems to lie in between determinsm and nondeterminism, with respect to space computations. See

H. R. Lewis, C.H. Papadimitriou (1982). Symmetric space-bounded computation. *Theoret. Comput. Sci. 19,* 161-187.

Whether the quadratic waste in Savitch's result can be improved upon is a major open problem. One of its versions, called the LBA problem for reasons too lengthy to explain here, enjoyed at some point the role played today by the $\mathcal{P}=?\mathcal{NP}$ puzzle. For work concerning another version, namely the question of whether SPACE($\log n$)$=?\mathcal{N}$SPACE($\log n$), see [Savitch 1970] above and

I.H. Sudborough (1975). On tape-bounded complexity classes and multihead finite automata. *J. Comput. System Sci. 10,* 62-76,

N.D. Jones, E. Lien, W.T. Laaser (1976). New problems complete for nondeterministic log space. *Math. Systems Theory 10,* 1-17.

The second major result relating classes of different sorts states that space is more powerful than time, in particular that TIME($F\log F$) is contained in SPACE(F); see

J.E. Hopcroft, W.J. Paul, L.G. Valiant (1975). On time versus space and related problems. *Proc. 16th Annual IEEE Symp. Foundations of Computer Science,* 57-64.

That proof employed an interesting *game* played on graphs, that seems to come up very frequently in modeling of computation. For a review of results on this game see

N. Pippenger (1981). Pebbling. *Proc. 5th IBM Symp. Mathematical Foundations of Computer Science.*

A very recent result shows that nondeterminism is more powerful than determinism! In particular, it is proven in

W.J. Paul, N. Pippenger, E. Szemeredi, W.T. Trotter (1983). On determinism versus nondeterminism and related problems. *Proc. 24th Annual IEEE Symp. Foundations of Computer Science,* 429-438,

that TIME(n) is strictly contained in $\mathfrak{N}$TIME(n). This is a difficult result, and it is the strongest result of this sort we have today. But in some sense it is a very weak result. To realize how poor our understanding still is in this direction, notice that we suspect that the latter class could also contain exponentially difficult problems. As we shall see in the next section, separating significantly determinism from nondeterminism is the central problem in complexity theory today.

Another important mode of computation could be called the *parallel* mode. In this mode, many machines cooperate in solving a problem, communicating somehow. This notion has become important with the advent of the 'supercomputer' with its phalanx of processors. See the bibliography by Kindervater and Lenstra in this volume. Defining what computation in this mode means is a formidable modeling problem, and many independent models of parallel computation have been shown to be all equivalent (more precisely: not to differ in power by more than a polynomial). For a few such examples see:

V.R. Pratt, L.J. Stockmeyer (1976). A characterization of the power of vector machines. *J. Comput. System Sci. 12,* 198-221,

W.J. Savitch (1978). Parallel and nondeterministic complexity classes. *Automata, Languages and Programming,* Springer, Berlin, 411-424,

A. Chandra, D.C. Kozen, L.J. Stockmeyer (1981). Alternation. *J. Assoc. Comput. Mach. 28,* 114-133,

S.A. Cook (1981). Towards a complexity theory of synchronous parallel computation. *Enseign. Math. (2) 27,* 99-124,

S. Fortune, J. Wyllie (1978). Parallelism in random access machines. *Proc. 10th Annual ACM Symp. Theory of Computing,* 114-119.

We know pretty much how this mode relates to determinism. The consensus in the above papers is that parallel time classes are as powerful as deterministic space classes with the same bound - give or take a polynomial. For more on parallel computational complexity see the bibliography by Kindervater and Lenstra in this volume.

A very different mode of computation (actually, a completely different model and rules) is what is known as *nonuniform computation.* Here we allow to have different algorithms for different input sizes. A natural example (and the main motivation) is a *circuit* consisting of gates and wires, capable of solving instances of a problem having a given length - say, eighty input bits. The resource could be the number of gates - this corresponds to deterministic time. However, because of the flexibility to change circuit with input size, nonuniform models could be more powerful than ordinary ones. For a taste of these issues see

R.M. Karp, R.J. Lipton (1980). Some connections between nonuniform and uniform complexity classes. *Proc. 12th Annual ACM Symp. Theory of Computing,* 302-309.

We end this section by recounting one of the earliest and most elegant achievements in the area, namely the *axiomatic approach to computational complexity.* It turns out that one can talk about complexity without specifying machine, resource, or mode of computation. An 'abstract resource' is a function defined for every problem and input, that has to satisfy two natural axioms. From these simple premises, a very interesting *theory of abstract complexity* (also known as *Blum complexity*) can be built:

M. Blum (1967). A machine-independent theory of the complexity of the recursive functions. *J. Assoc. Comput. Mach. 14,* 322-336,

E.M. McCreight, A.R. Meyer (1969). Classes of computable functions defined by bounds on computation. *Proc. 1st Annual ACM Symp. Theory of Computing,* 79-88.

For an exposition of abstract complexity see [Machtey & Young 1978] (§2) and

J. Hartmanis, J.E. Hopcroft (1971). An overview of the theory of computational complexity. *J. Assoc. Comput. Math. 18,* 444-475.

The major results of this theory, which borrows form *recursion theory* in methodology and style, state that complexity classes exhibit some very counterintuitive behavior. The root of those anomalies is that the complexity bounds are not required to be constructible, as we have required in this exposition.

Finally, for a nice overview of computational complexity circa a decade ago (when its basic character was starting to take shape) see

A. Borodin (1973). Computational complexity: theory and practice. A.V. Aho (ed.). *Currents in the Theory of Computing,* Prentice-Hall, Englewood Cliffs, NJ, 35-89.

4. $\mathcal{P}$, $\mathcal{NP}$, AND VICINITY

In 1971 Cook described what would become the central problem in computational complexity, namely whether TIME(n^c)=$\mathcal{NTIME}$(n^c) (or $\mathcal{P}=\mathcal{NP}$, as these important classes are now denoted). He also presented the first $\mathcal{NP}$-complete problem, namely the satisfiability of Boolean expressions:

S.A. Cook (1971). The complexity of theorem-proving procedures. *Proc. 3rd Annual ACM Symp. Theory of Computing,* 151-158.

Karp established the importance of these ideas by exhibiting a number of $\mathcal{NP}$-complete problems, most of them originating from optimization:

R.M. Karp (1972). Reducibility among combinatorial problems. R.E. Miller, J.W. Thatcher (eds.). *Complexity of Computer Computations,* Plenum, New York, 85-103.

Similar investigations were carried out around the same time in the Soviet Union; see

L.A. Levin (1973). Universal sorting problems. *Problemi Peredachi Informatsii 9,* 115-116; English translation: *Problems Inform. Transmission 9,* 265-266.

The long wait for a proof that $\mathcal{P} \neq \mathcal{NP}$ has not slowed down the field at all. The concept of $\mathcal{NP}$-completeness has been an important and popular criterion for classifying problems, even in the absence of concrete proof that it indeed implies intractability. The encyclopedia of the subject is of course the book by [Garey & Johnson 1979] (see §1), where, besides an impressively thorough sensus of $\mathcal{NP}$-complete problems, there is a nice introduction to complexity and the methodology of proving problems $\mathcal{NP}$-complete. What is more, the book has become a serial:

D.S. Johnson (1981-). The NP-completeness column: an ongoing guide. *J. Algorithms.*

However, the question of whether $\mathcal{P}=\mathcal{NP}$ did not only give us the concept of $\mathcal{NP}$-completeness, but also a whole treasure of surrounding classes and issues. Problems in $\mathcal{NP}$ have *good characterizations* according to Edmonds, that is, they can be obtained from polynomial problems, with the prefix of existential quantifiers. For example, satisfiability can be obtained from the problem of determining whether a truth assignment satisfies a Boolean formula (clearly in $\mathcal{P}$) by existentially quantifying away the assignment. Universal quantifiers give us the class co$\mathcal{NP}$, of problems that are complements of problems in $\mathcal{NP}$. Why stop? Since adding more quantifiers of the same kind doesn't buy us anything new, let us alternate quantifiers. We obtain an infinite hierarchy of complexity classes, called the *polynomial hierarchy*; see

L.J. Stockmeyer (1976). The polynomial-time hierarchy. *Theoret. Comput. Sci. 3,* 1-22.

An interesting result is that by allowing infinite alternations we obtain SPACE(n^c), or $\mathcal{P}$SPACE:

L.J. Stockmeyer, A.R. Meyer (1973). Word problems requiring exponential space. *Proc. 5th Annual ACM Symp. Theory of Computing,* 1-9.

Notice that, due to the theorem by Savitch, $\mathcal{P}$SPACE$=\mathcal{NP}$SPACE. $\mathcal{P}$SPACE contains $\mathcal{P}$, $\mathcal{NP}$, and the polynomial hierarchy, but we don't know whether the inclusions are proper. So, for all we know, the polynomial hierarchy might not be. There are several $\mathcal{P}$SPACE-complete problems; for the principal kinds see

T.J. Schaefer (1978). Complexity of some two-person complete information games. *J. Comput. System Sci. 16,* 185-225,

A.S. Fraenkel, M.R. Garey, D.S. Johnson, T. Schaefer, Y. Yesha (1978). The complexity of checkers on an $N \times N$ board. *Proc. 19th Annual IEEE Symp. Foundations of Computer Science,* 55-64,

J. Orlin (1981). The complexity of dynamic languages and dynamic optimization problems. *Proc 13th Annual ACM Symp. Theory of Computing,* 218-227,

C.H. Papadimitriou (1983). Games against nature. *Proc. 24th Annual IEEE Symp. Foundations of Computer Science,* 446-450.

Interestingly, the two latter papers introduce $\mathcal{P}$SPACE-complete problems from optimization (namely, periodic optimization, and decision-making under uncertainty, respectively). Certain complete problems for classes lower in the hierarchy (in fact, between $\mathcal{NP}$ and the second level) also have an optimization flavor:

C.H. Papadimitriou, M. Yannakakis (1982). The complexity of facets (and some facets of complexity). *Proc 14th Annual ACM Symp. Theory of Computing,* 255-260,

C.H. Papadimitriou (1984). On the complexity of unique solutions. *J. Assoc. Comput. Mach. 31,* 392-400.

Also 'between' $\mathcal{NP}$ and $\mathcal{P}$SPACE lies the class $\#\mathcal{P}$ of *enumeration problems*, again of interest to combinatorial optimization. See

L.G. Valiant (1979). The complexity of computing the permanent. *Theoret. Comput. Sci. 8,* 189-201,

L.G. Valiant (1979). The complexity of enumeration and reliability problems. *SIAM J. Comput. 8,* 410-421.

In fact, there have been classes of problems introduced that are 'between' $\mathcal{P}$ and $\mathcal{NP}$. The motivation was to define more relaxed notions of 'feasible computation'. A very important such attempt is the concept of *randomized polynomial computation*, carried out in machines that can take random steps. This is distinct from the *probabilistic* approach to complexity, reviewed in this volume by Karp, Lenstra, McDiarmid and Rinnooy Kan, in which the behavior of deterministic algorithms is analyzed probabilistically, by assuming certain distributions on the inputs. For the randomized approach, see the bibliography by Maffioli, Speranza and Vercellis in this volume, and

J. Gill (1976). Complexity of probabilistic Turing machines. *SIAM J. Comput. 6,* 675-695,

M.O. Rabin (1976). Probabilistic algorithms. J.F. Traub (ed.). *Algorithms and Complexity: New Directions and Recent Results,* Academic Press, New York, 21-39,

L. Adleman (1978). Two theorems in random polynomial time. *Proc. 19th Annual IEEE Symp. Foundations of Computer Science,* 75-83,

M. Sipser (1983). A complexity-theoretic approach to randomness. *Proc. 15th Annual ACM Symp. Theory of Computing,* 330-335.

In relation to $\mathcal{P}$ and $\mathcal{NP}$, an interesting concept is that of *relativization*. Using this concept, it has been shown, for example, that there are 'universes' in which $\mathcal{P}=\mathcal{NP}$, whereas in other 'universes' the opposite holds. A 'universe' is, roughly speaking, an environment in which certain computations are free. Our universe, of course, has no free computations. Such results, however, help demonstrate the difficulty of the $\mathcal{P}=?\mathcal{NP}$ problem, since they suggest that it cannot be settled by techniques that transcend 'universes' - and most known techniques for proving upper and lower bounds do. For such results see

T. Baker, J. Gill, R. Solovay (1975). Relativizations of the P=?NP question. *SIAM J. Comput. 4,* 431-442,

T. Baker, A. Selman (1976). A second step towards the polynomial hierarchy. *Proc. 17th Annual IEEE Symp. Foundations of Computer Science,* 71-75,

C.H. Bennett, J. Gill (1981). Relative to a random oracle A, $P^A \neq NP^A \neq$ co-NP^A with probability 1. *SIAM J. Comput. 10,* 96-113.

These ideas have their origins in *recursion theory*. Another interesting result that uses recursive-theoretic techniques is the one stating that, if $\mathcal{P} \neq \mathcal{NP}$, then there are problems 'in between', which are not $\mathcal{NP}$-complete, but are not in $\mathcal{P}$ either:

R.E. Ladner (1975). On the structure of polynomial-time reducibility. *J. Assoc. Comput. Math. 22,* 155-171.

If a problem is hard, how frequent are its difficult instances? This raises the question of *sparsity* of hard problems. For the latest on that see

J. Hartmanis, V. Sewelson, N. Immerman (1983). Sparse Sets in $NP-P$: EXPTIME vs. NEXPTIME. *Proc 15th Annual ACM Symp. Theory of Computing,* 382-391.

One cannot talk about the $\mathcal{P}=?\mathcal{NP}$ problem and research on it in the last decade without mentioning a few important problems for which difficult positive results were shown recently. These include *linear programming*:

L.G. Khachian (1979). A polynomial algorithm in linear programming. *Soviet Math. Dokl. 20,* 191-194;

graph isomorphism:

E.M. Luks (1980). Isomorphism of graphs of bounded valence can be tested in polynomial time. *Proc. 21st Annual IEEE Symp. Foundations of Computer Science,* 42-49;

integer programming with bounded number of variables:

H.W. Lenstra, Jr. (1983). Integer programming with a fixed number of variables. *Math. Oper. Res. 8,* 538-548;

matroid parity:

L. Lovász (1980). Matroid matching and some applications. *J. Combin. Theory Ser. B 28,* 121-131;

and *prime number recognition*, see [Rabin 1976] above and

G.L. Miller (1976). Riemann's hypothesis and tests for primality. *J. Comput. System Sci. 13,* 300-317,

L.M. Adleman (1980). On distinguishing prime numbers from composite numbers (abstract). *Proc. 21st Annual IEEE Symp. Foundations of Computer Science,* 387-406.

What is most exciting about these results is that they make nontrivial use of facts and techniques from other mathematical disciplines such as *number theory* and the *geometry of numbers, group theory, nonlinear programming*, and so on. They succeed in inspiring some hope that not all difficult problems are computationally intractable. Such optimism is not considered absolutely absurd, even in the context of $\mathcal{P}$ and $\mathcal{NP}$. On the other hand, there have been some interesting exponential lower bounds for restricted models of computation, reminding us of the Holy Grail of this field, a proof that $\mathcal{P}$ is not $\mathcal{NP}$, and how far we still are from it:

C. McDiarmid (1976). *Determining the Chromatic Number of a Graph,* Report STAN-CS-76-576, Computer Science Department, Stanford University,

V. Chvátal (1977). Determining the stabilility number of a graph. *SIAM J. Comput. 6,* 643-662,

M. Furst, J.B. Saxe, M. Sipser (1981). Parity, circuits, and the polynomial-time hierarchy. *Proc 22nd Annual IEEE Symp. Foundations of Computer Science,* 260-270.

6

Probabilistic Analysis

R.M. Karp
University of California, Berkeley

J.K. Lenstra
Centre for Mathematics and Computer Science, Amsterdam

C.J.H. McDiarmid
Wolfson College, Oxford

A.H.G. Rinnooy Kan
Erasmus University, Rotterdam

CONTENTS

The analysis of combinatorial algorithms is traditionally concerned with *worst case* time and space bounds. Such an analysis has to account for the isolated time consuming problem instance, and hence the results may be pessimistic and give a misleading picture of the *average case.* This point is supported by an abundance of empirical evidence. Thus the ultimate explanation of why algorithms behave as they do must be of a *probabilistic* nature.

A probabilistic analysis requires first of all the specification of a probability distribution over the set of all problem instances. For example, several models for generating random graphs have been extensively investigated, but for other combinatorial structures the choice of a reasonable probability model is less obvious.

A probabilistic analysis of combinatorial problems and algorithms is usually far from trivial. The main reasons for this are the discrete structure of problem instances and solutions, as well as the interdependence between the various steps of an algorithm. What happens at a nodc of a search tree, for example, depends highly on what has happened at its predecessor.

In recent years, progress has been made on various fronts. One of these is *probabilistic running time analysis.* An example of this approach is the collective effort to explain the success of the simplex method for linear programming. One of the great challenges here is to give rigorous proofs of the polynomial expected running time of various search algorithms, in order to confirm informal analyses or empirical evidence. Secondly, there is the area of *probabilistic error analysis,* where the error refers to the (absolute or relative) difference between an approximate solution value and the optimum. The empirical behavior of heuristics suggests that the worst case error is seldom met in practice, but analytical verification may be quite difficult. Much research of this type is actually based on *probabilistic value analysis,* the third and perhaps most surprising area. Many hard combinatorial optimization problems, especially those with a Euclidean structure, allow a simple probabilistic description of their optimal solution value in terms of the problem parameters.

This bibliography concentrates on these types of probabilistic analyses in combinatorial optimization. It excludes other approaches involving probability models, notably *randomized algorithms* (see Ch.7) and *stochastic optimization* problems in which the realization of the data is not known in advance (see, e.g., Ch.11, §11.2 on stochastic scheduling). We have also excluded topics that are insufficiently related to the area of combinatorial optimization, such as probabilistic models for sorting and for VLSI circuit design.

The organization of this bibliography is as follows. §1 lists a number of *surveys* on the probabilistic analysis of combinatorial algorithms. §2 presents the *basic tools* that are used in the area; the references selected here are intended only as a means of access to the literature on this subject. §3 and §4 review results on *unweighted* and *weighted graphs,* respectively. Most papers in §3 deal with the problems of finding matchings, stable sets, colorings, and Hamiltonian cycles in random graphs. The main subjects in §4 are the problems of finding optimal assignments and shortest traveling salesman tours. §5 is concerned with problems defined in *Euclidean space,* with emphasis on location and routing problems. §6 surveys the recent literature on the average case behavior of variants of the *simplex method* for linear programming. §7 collects results on other *non-graphical, non-Euclidean* problems, such as bin packing, scheduling and knapsack problems. §8 finally discusses *search techniques* for solving hard problems.

This research was partially supported by NSF grants MCS-8105217 and MCS-8311422.

1. Surveys

R.M. Karp (1976). The probabilistic analysis of some combinatorial search algorithms. J.F. Traub (ed.). *Algorithms and Complexity: New Directions and Recent Results,* Academic Press, New York, 1-19.

A general framework for the probabilistic analysis of combinatorial algorithms is introduced. Several algorithms are analyzed probabilistically, including a cellular dissection algorithm for the Euclidean traveling salesman problem, sequential algorithms for constructing cliques and colorings, an extension-rotation algorithm for the Hamiltonian cycle problem and a tree search algorithm for the approximate solution of set covering problems.

G. d'Atri, C. Puech (1978). Analyse probabilistique des problèmes combinatoires. *Mathématiques Appliquées, 1er Coll. AFCET-SMF, Tome II,* Palaiseau, 261-273.

A survey of probabilistic approaches to the analysis of combinatorial problems. The principal examples concern random graphs and random shortest path problems.

G. d'Atri (1980). Outline of a probabilistic framework for combinatorial optimization. F. Archetti, M. Cugiani (eds.). *Numerical Techniques for Stochastic Systems,* North-Holland, Amsterdam, 347-368.

The basic concepts of the probabilistic analysis of algorithms and of the concept of a randomized algorithm are explained through examples: a search problem, the knapsack problem, and primality testing.

L. Slominski (1982). Probabilistic analysis of combinatorial algorithms: a bibliography with selected annotations. *Computing 28,* 257-267.

An excellent survey with especially good coverage of early Soviet work not available in English. This survey is a useful complement to the present one, which covers the early Soviet literature rather sparsely.

D.S. Johnson (1984). The NP-completeness column: an ongoing guide; eleventh edition. *J. Algorithms 5,* 284-299.

This column first surveys results that show $\mathcal{NP}$-complete problems to be solvable in polynomial time on average. It then considers the concept of 'hardness on average' and discusses an important result of L.A. Levin (see §7.1).

R.M. Karp (to appear). The probabilistic analysis of combinatorial optimization algorithms. *Proc. Internat. Congress Math.,* Warsaw.

Several problems are discussed from the viewpoint of probabilistic analysis. These include multiprocessor scheduling, matchings and Hamiltonian cycles, cliques and colorings, the assignment problem and the traveling salesman problem.

2. Basic Tools

2.1. *Moment methods*

P. Erdös, J. Spencer (1974). *Probabilistic Methods in Combinatorics,* Academic Press, New York.
B. Bollobás (1979). *Graph Theory,* Graduate Texts in Mathematics 63, Springer, Berlin.

Existence results concerning random graphs or networks are often proved by using a *moment method.* Let the random variable X take values 0,1,2,..., and have mean $E[X] = \mu > 0$ and variance $E[(X-\mu)^2] = \sigma^2$. (Perhaps X counts the number of subgraphs of a random graph with a certain property.) Then we have
(a) the *Markov Inequality*: $Pr\{X>0\} \leqslant \mu$;
(b) from the *Chebyshev Inequality*: $Pr\{X = 0\} \leqslant \sigma^2/\mu^2$;
(c) from the *Cauchy-Schwarz Inequality*: $Pr\{X>0\} \geqslant \mu^2/E[X^2]$.
When we use (a) [(b) or (c)], then we say that we are using a first [second] moment method.

2.2. *Binomial properties*

W. Feller (1968). *An Introduction to Probability Theory and Its Applications, Volume 1, Third Edition,* Wiley, New York.
D.S. Mitrinovic (1970). *Analytic Inequalities,* Springer, Berlin.

Let $0 < p = 1-q < 1$ and let n be a positive integer. A key result in combinatorial probability is *Stirling's Formula* (as refined by H. Robbins):

$$n! = n^n e^{-n} \sqrt{2\pi n}\, e^{\alpha(n)}, \text{ where } \frac{1}{12n+1} < \alpha(n) < \frac{1}{12n}.$$

The following simple bounds for binomial coefficients are often useful. If $1 \leqslant k \leqslant n$ then

$$\left(\frac{n}{k}\right)^k \leqslant \binom{n}{k} < \left(\frac{ne}{k}\right)^k.$$

There exists a constant $c>0$ (independent of p) such that, if np is an integer, then

$$\binom{n}{np} p^{np} q^{nq} \geqslant cn^{-\frac{1}{2}}.$$

2.3. *Bounds on tails of distributions*

H. Chernoff (1952). A measure of asymptotic efficiency for tests of a hypothesis based on a sum of observations. *Ann. Math. Statist. 23,* 493-507.

Chernoff proves the following bounds on the tails of the distribution of the sum of independent observations. Let $X_1,...,X_n$ be independent identically distributed random variables with finite expectation μ and let $m(\alpha) = \inf_t \{e^{-\alpha t} E[e^{tX_1}]\}$. If $\alpha \leqslant [\geqslant] \mu$, then $Pr\{\Sigma_{k=1}^{n} X_k \leqslant [\geqslant] n\alpha\} \leqslant (m(\alpha))^n$. For the tails of a binomial distribution this implies that, if $0 \leqslant \beta \leqslant 1$, then

$$\sum_{k \leqslant (1-\beta)np} \binom{n}{k} p^k q^{n-k} \leqslant e^{-\beta^2 np/2},$$

$$\sum_{k \geqslant (1+\beta)np} \binom{n}{k} p^k q^{n-k} \leqslant e^{-\beta^2 np/3}.$$

These inequalities as well as the last inequality in §2.2 were used, for example, in [Angluin & Valiant 1979] (see §3.3, §3.7).

W. Hoeffding (1963). Probability inequalities for sums of bounded random variables. *J. Amer. Statist. Assoc. 58,* 13-30.
V. Chvátal (1979). The tail of the hypergeometric distribution. *Discrete Math. 25,* 285-287.

Related results and extensions.

A.W. Marshall, I. Olkin (1979). *Inequalities: Theory of Majorisation and Its Applications,* Academic Press, New York.

Chapter 17C gives extensions of the following intuitively obvious result that has on occasion been useful. Let $X_1, X_2, ...$ be a sequence of 0,1-valued random variables such that for $t = 1,2,...$, given any history concerning $X_1,...,X_t$, the probability that $X_{t+1} = 1$ is at most p. Then $X_1 + ... + X_n$ is stochastically less than a binomial random variable with parameters n and p.

P.J. Boland, F. Proschan (1983). The reliability of k out of n systems. *Ann. Probab. 11,* 760-764.

A related result of Hoeffding set in the context of majorization.

2.4. *Conditioning*

M.L. Eaton (1982). A review of selected topics in multivariate probability inequalities. *Ann. Statist. 10,* 11-43.
R.L. Graham (1983). Applications of the FKG inequality and its relatives. A. Bachem, M. Grötschel, B. Korte (eds.). *Mathematical Programming: the State of the Art - Bonn 1982,* Springer, Berlin, 115-131.

Two recent reviews of the most useful results on 'benevolent' conditioning: *Harris' Lemma* and its extension, the *FKG Inequality.*

Y.L. Tong (1980). *Probability Inequalities in Multivariate Distributions,*

Academic Press, New York.

A useful concept of 'positive dependence' is that of *association* - see for example [Eaton 1982] above and this reference.

K. Joag-Dev, F. Proschan (1983). Negative association of random variables, with applications. *Ann. Statist. 11,* 286-295.

A recent paper on one of the various concepts of 'negative dependence' that have been proposed.

C. McDiarmid (1981). General percolation and random graphs. *Adv. in Appl. Probab. 13,* 40-60.
C. McDiarmid (1983). General first-passage percolation. *Adv. in Appl. Probab. 15,* 149-161.

Certain 'general percolation' results are useful for handling random directed graphs and networks. For example, the random directed graph $D_{n,p}$ is more likely than the random graph $G_{n,p}$ to have a Hamiltonian cycle.

C. McDiarmid (to appear). On some conditioning results in the probabilistic analysis of algorithms. *Discrete Appl. Math.*

A simple combinatorial approach for handling conditioning problems that arise in the probabilistic analysis of graph algorithms. Arguments from [Angluin & Valiant 1979] (see §3.3, §3.7) and [Karp & Tarjan 1980] (see § 3.2) are substantially simplified.

2.5. *Stochastic convergence*

R.J. Serfling (1980). *Approximation Theorems of Mathematical Statistics,* Wiley, New York.

A sequence of random variables $X_1, X_2, \ldots$ is said to converge to a random variable X
(a) *in probability* if $\lim_{n \to \infty} Pr\{|X_n - X| > \epsilon\} = 0$ for every $\epsilon > 0$;
(b) *with probability* 1 or *almost surely* ('a.s.') if $Pr\{\lim_{n \to \infty} X_n = X\} = 1$;
(c) *completely* if $\sum_{n=1}^{\infty} Pr\{|X_n - X| > \epsilon\} < \infty$ for every $\epsilon > 0$.
According to the *Borel-Cantelli Lemma,* (c) implies (b). Also, (b) implies (a), but the inverse implications do not hold. For comments on these concepts that are relevant in the present context, see [Karp & Steele 1985, §2.4] in §4.5.

3. Unweighted Graphs

3.1. *Random graphs*

In this section we follow the loose but common practice of saying that an event concerning random graphs happens *almost surely* or for *almost all* graphs if the probability that it happens tends to 1 as $n \to \infty$.

B. Bollobás (to appear). *Lectures on Random Graphs.*

Let $G_{n,p}$ $[D_{n,p}]$ denote the random graph [directed graph] with vertex set $\{1, \ldots, n\}$ in which the $n(n-1)/2$ $[n(n-1)]$ possible edges occur independently with probability p. The random graph $G_{n,N}$ [directed graph $D_{n,N}$] has the same vertex set but now the $\binom{n(n-1)/2}{N}$ possible graphs $[\binom{n(n-1)}{N}$ possible directed graphs] occur with the same probability.

[Erdös & Spencer 1974] and [Bollobás 1979] (see §2.1) give an introduction to the theory of random graphs. The current reference gives a full treatment. For relations between the models $G_{n,p}$ and $G_{n,N}$ see, for example, [Angluin & Valiant 1979] (§3.3, §3.7).

B. Bollobás (1981). Random graphs. H.N.V. Temperley (ed.). *Combinatorics,* London Mathematical Society Lecture Notes 52, 80-102.
M. Karónski (1982). A review of random graphs. *J. Graph Theory 6,* 349-389.
K. Weber (1982). Random graphs - a survey. *Rostock. Math. Kolloq. 21,* 83-98.
G. Grimmett (1983). Random graphs. L. Beineke, R. Wilson (eds.). *Selected Topics in Graph Theory 2,* Academic Press, London, 201-235.

Four recent surveys on random graphs. For a discussion of random directed graphs see [McDiarmid 1981] (§2.4).

3.2. *Connectivity*

We consider here some problems related to the connectivity of a graph or directed graph for which there exist algorithms that are quite fast in the worst case.

P.A. Bloniarz, M.J. Fischer, A.R. Meyer (1976). A note on the average time to compute transitive closures. S. Michaelson, R. Milner (eds.). *Automata, Languages and Programming,* Edinburgh University Press, Edinburgh, 425-434.

An algorithm for the transitive closure of a directed graph has average time $O(n^2 \log n)$. The analysis is for random directed graphs with probabilities depending only on the number of vertices and set of outdegrees (more general than $D_{n,N}$).

C.P. Schnorr (1978). An algorithm for transitive closure with linear expected time. *SIAM J. Comput. 7,* 127-133.

An algorithm for transitive closure is given with average time $O(n+m^*)$ where n is the number of vertices and m^* is the expected number of edges in the transitive closure. The analysis is for random directed graphs $D_{n,N}$.

R.M. Karp, R.E. Tarjan (1980). Linear expected time algorithms for connectivity problems. *J. Algorithms 1,* 374-393.

Algorithms that run in linear expected time are given for finding connected components, strong components and biconnected components. The analysis is for random graphs $G_{n,N}$ and random directed graphs $D_{n,N}$ (uniformly over N).

3.3. *Matching*

A *matching* in a graph is a set of edges with no endpoints in common. In a graph with n vertices, a *maximum* matching (a matching of maximum cardinality) can be found in time $O(n^{2.5})$; it is *perfect* if it contains $n/2$ edges. A *cover* of vertices by edges is a set of edges such that each vertex is an endpoint of some edge in the set. A minimum cover consists of a maximum matching together with any minimal set of edges covering the remaining vertices. For some early papers on matchings and coverings see [Slominski 1982] (§1).

D. Angluin, L.G. Valiant (1979). Fast probabilistic algorithms for Hamiltonian circuits and matchings. *J. Comput. System Sci. 19,* 155-193.

An $O(n \log n)$ time heuristic a.s. finds a perfect matching in $G_{n,N}$ when n is even and $N \geqslant cn \log n$, for a suitable constant c. Earlier related work is discussed.

W.F. de la Vega (1980). Sur la cardinalité maximum des couplages d'hypergraphes aléatoires uniformes. *Discrete Math. 40,* 315-318.

If $N/n \to \infty$ as $n \to \infty$, then a greedy heuristic yields a matching M_n in $G_{n,N}$ with $|M_n|/n \to ½$ in probability; whilst with fewer edges the proportion of isolated vertices does not tend to 0 in probability.

R.M. Karp, M. Sipser (1981). Maximum matchings in sparse random graphs. *Proc. 22nd Annual IEEE Symp. Foundations of Computer Science,* 364-375.

A linear time heuristic based on trimming away low-degree vertices is shown to give near maximum matchings in $G_{n,p}$ when $p = \lambda/(n-1)$.

E. Shamir, E. Upfal (1981). On factors in random graphs. *Israel J. Math. 39,* 296-302.

An existence result concerning f-factors in $G_{n,p}$ is proved by showing that subfactors can almost surely be augmented by using alternating paths.

E. Shamir, E. Upfal (1982). N-Processors graphs distributively achieve perfect matchings in $O(\log^2 N)$ beats. *Proc. Annual ACM Symp. Principles of Distributed Computing,* 238-241.

A parallel algorithm (one processor at each vertex, no shared memory) operates for $O(\log^2 n)$ beats and a.s. finds a perfect matching in $G_{n,p}$ when $np > c \log n$ (for a suitable constant c).

G. Tinhofer (1984). A probabilistic analysis of the greedy heuristic for the matching problem. *Ann. Oper. Res. 1.*

Theoretical and simulation results are given concerning variants of a greedy heuristic for maximum matchings.

A.M. Frieze (1984A). *On Large Matchings and Cycles in Sparse Random*

Graphs, Graduate School of Industrial Administration, Carnegie-Mellon University, Pittsburgh, PA.

The random graph $G_{n,c/n}$ a.s. contains a matching of cardinality $n(1-(1+\epsilon(c))e^{-c})/2$, where $\epsilon(c)\to 0$ as $c\to\infty$.

A.M. Frieze (1984B). *Maximum Matchings in a Class of Random Graphs,* Graduate School of Industrial Administration, Carnegie-Mellon University, Pittsburgh, PA.

Let G_n^m denote the random graph with n vertices in which each vertex independently chooses m edges incident with it. G_n^1 a.s. does not contain a perfect matching, and G_n^m ($m \geq 2$, n even) a.s. does.

3.4. *Graph isomorphism*

It is not known if the problem of testing whether two graphs are isomorphic is $\mathcal{NP}$-complete, although if the graphs have bounded degree the problem is known to be in $\mathcal{P}$.

R.J. Lipton (1978). *The Beacon Set Approach to Graph Isomorphism,* Yale University.
L. Babai, L. Kučera (1979). Canonical labelling of graphs in linear average time. *Proc. 20th Annual IEEE Symp. Foundations of Computer Science,* 39-46.
R.M. Karp (1979). Probabilistic analysis of a canonical numbering algorithm for graphs. *Proc. Symp. Pure Mathematics 34,* AMS, Providence, RI, 365-378.
L. Babai, P. Erdös, S.M. Selkow (1980). Random graph isomorphism. *SIAM J. Comput. 9,* 628-635.

Each of these papers gives a fast *canonical labeling* algorithm that works for almost all graphs. Thus a.s. for $G_{n,1/2}$ any graph can be tested for isomorphism to this graph by a naive fast algorithm. The paper listed last above discusses the papers listed earlier. See also [Johnson 1984] (§1).

3.5. *Stable sets and coloring*

A set of vertices in a graph G is *stable* if no two are adjacent. The *stability number* $\alpha(G)$ is the maximum size of a stable set in G. A *coloring* of G is an assignment of colors to the vertices so that no two adjacent vertices receive the same color. The *chromatic number* $\chi(G)$ is the least number of colors in a coloring of G.

No polynomial time algorithms are known to approximate either $\alpha(G)$ or $\chi(G)$ to within any constant factor. If $\mathcal{P}\neq\mathcal{NP}$ then no such algorithm colors within a ratio less than 2.

In [Erdös & Spencer 1974] (see §2.1) bounds are stated for $\chi(G_{n,p})$ both in the *constant density* case (Ch. 11, Exercise 2) and in the *constant average degree* case (Ch. 16, Exercise 9).

C. McDiarmid (1984). Colouring random graphs. *Ann. Oper. Res.* 1.

The greedy stable set algorithm considers the vertices in a given order and adds them to the current stable set if possible. The greedy or simple sequential coloring algorithm does this repeatedly to form the different color sets. McDiarmid's survey (with 57 references) focuses mainly on approaches of this type. It includes a treatment of the constant average degree case.

G. Grimmett, C. McDiarmid (1975). On colouring random graphs. *Math. Proc. Cambridge Philos. Soc. 77,* 313-324.

In the constant density random graph $G_{n,p}$ simple greedy approaches a.s. yield a stable set of size at least $(\frac{1}{2}-\epsilon)\alpha(G_{n,p})$ and a coloring using at most $(2+\epsilon)\chi(G_{n,p})$ colors.

B. Bollobás, P. Erdös (1976). Cliques in random graphs. *Math. Proc. Cambridge Philos. Soc. 80,* 419-427.
D.W. Matula (1976). *The Largest Clique Size in a Random Graph,* Technical report CS7608, Department of Computer Science, Southern Methodist University, Dallas, TX.

Results from [Grimmett & McDiarmid 1975] (see above) are sharpened.

V. Chvátal (1977). Determining the stability number of a graph. *SIAM J. Comput. 6,* 643-662.

For almost all random graphs with (large) constant average degree all recursive proofs bounding the stability number are of exponential length, and hence any 'Tarjan-type' algorithm must be slow.

L. Kučera (1977). Expected behaviour of graph coloring algorithms. M. Karpinski (ed.) *Fundamentals of Computation Theory,* Lecture Notes in Computer Science 56, Springer, Berlin, 447-451.

The expected behavior of the simple greedy coloring algorithm and of a variant (involving considering vertex degrees) are discussed when they act on random graphs $G_{n,p}$ and on random k-partite graphs.

C. McDiarmid (1979). Determining the chromatic number of a graph. *SIAM J. Comput. 8,* 1-14.

For constant density random graphs $G_{n,p}$ a.s. all algorithms in a certain class of branch-and-bound algorithms for determining the chromatic number will take more than exponential time.

C. McDiarmid (1979). Colouring random graphs badly. R.J. Wilson (ed.). *Graph Theory and Combinatorics,* Pitman Research Notes in Mathematics 34, Pitman, London, 76-86.

For constant density random graphs $G_{n,p}$ the greedy or simple sequential coloring algorithm a.s. performs essentially as badly as possible. Results on the achromatic number are tightened up later in [McDiarmid 1982] (see below).

A.D. Korsunov (1980). The chromatic number of n-vertex graphs. *Metody Diskret. Analiz. 35,* 14-44, 104 (in Russian).

The conjecture that $\chi(G_{n,\frac{1}{2}})\,(\log_2 n)/n \rightarrow \frac{1}{2}$ in probability as $n \rightarrow \infty$ is given as a theorem. The proof uses the second moment method.

C. McDiarmid (1982). Achromatic numbers of random graphs. *Math. Proc. Cambridge Philos. Soc. 92,* 21-28.

The *achromatic number* $\psi(G)$ of a graph G is the largest number of colors in a coloring of G such that no two colors may be identified. For constant density random graphs $G_{n,p}$, it is shown that a.s. $n/(k+1) \leqslant \psi(G_{n,p}) \leqslant n/(k-1)$, where $k = \log n / \log(1/(1-p))$. The lower bound is obtained by an analysis of a silly variant of the greedy coloring algorithm.

B. Pittel (1982). On the probable behaviour of some algorithms for finding the stability number of a graph. *Math. Proc. Cambridge Philos. Soc. 92,* 511-526.

This paper uses martingale arguments to analyze the greedy stable set algorithm, and investigates Chvátal's 'f-driven' algorithms for determining the stability number (without subtleties of the monotone rule).

A. Johri, D.W. Matula (1982). *Probabilistic Bounds and Heuristic Algorithms for Coloring Large Random Graphs,* Technical report 82-CSE-6, Southern Methodist University, Dallas, TX.

Nonasymptotic theoretical work together with simulation results indicate that with high probability the random graph $G_{1000,\frac{1}{2}}$ has chromatic number in the range 85 ± 12.

T. Kawaguchi, H. Nakano, Y. Nakanishi (1982). *Probabilistic Analysis of a Heuristic Graph Colouring Algorithm,* Unpublished manuscript.

Asymptotic results are given for the greedy coloring algorithm acting on $G_{n,p}$ when $np = cn^{\delta}$ for some constants c and $\frac{1}{2} < \delta \leqslant 1$. Stronger results appear in [McDiarmid 1983], [De la Vega 1982A] and [Shamir & Upfal 1984] (see below). Some nonasymptotic results are also given.

W.F. de la Vega (1982A). *On the Chromatic Number of Sparse Random Graphs,* Laboratoire de Recherche en Informatique, Université de Paris-Sud.
E. Shamir, E. Upfal (1984). Sequential and distributed graph coloring algorithms with performance analyses in random graph spaces. *J. Algorithms.*

Both these papers consider random graphs $G_{n,p}$ with decreasing density and increasing average degree, more specifically $p \rightarrow 0$ and $np \rightarrow \infty$ as $n \rightarrow \infty$. They show that a variant of the greedy coloring algorithm (involving a different end phase) is a.s. optimal to within a factor $2+\epsilon$, and thus settle a conjecture of Erdös and Spencer.

W.F. de la Vega (1982B). *Crowded Graphs Can Be Colored Within a Factor $1+\epsilon$ in Polynomial Time,* Laboratoire de Recherche en Informatique,

Université de Paris-Sud.

Consider very dense random graphs for which $\alpha(G_{n,p})$ is a.s. a constant $r+1$. A greedy heuristic for picking disjoint stable sets of size r is a.s. optimal to within a factor $1+\epsilon$.

C. McDiarmid (1983). On the chromatic forcing number of a random graph. *Discrete Appl. Math. 5,* 123-132.

If we wish to compute lower bounds for the chromatic number $\chi(G)$ of a graph G we may be interested in the *chromatic forcing number* $f_\chi(G)$ which is defined to be the least number of vertices in a subgraph H of G with $\chi(H) = \chi(G)$. For random graphs $G_{n,p}$ with say $(\log n)^{-1} < p(n) < 1-(\log n)^{-1}$ we have $f_\chi \geqslant (\frac{1}{2}-\epsilon)n$ a.s.

D.W. Matula (1983). *Improved Bounds on the Chromatic Number of a Graph,* Abstract, Department of Computer Science, Southern Methodist University, Dallas, TX.

A certain nonpolynomial time coloring algorithm is a.s. optimal to within a factor $(3/2+\epsilon)$, thus beating the greedy coloring algorithm.

H.S. Wilf (1984). Backtrack: an $O(1)$ expected time algorithm for the graph coloring problem. *Inform. Process. Lett. 18,* 119-121.
E.A. Bender, H.S. Wilf (1984). A theoretical analysis of backtracking in the graph coloring problem. *J. Algorithms.*

A simple backtracking procedure will test if a graph can be colored with k colors. For fixed k, the average time taken for certain random graphs $G_{n,p}$ is shown to be small. See also [Johnson 1984] (§1).

We conclude this subsection with results for miscellaneous problems related to finding stable sets and colorings in graphs.

G. Cornuéjols, G.L. Nemhauser, L.A. Wolsey (1978). *Worst-Case and Probabilistic Analysis of Algorithms for a Location Problem,* Technical report 375, School of Operations Research and Industrial Engineering, Cornell University, Ithaca, NY.

The problem considered here is to choose a set S of k vertices in a given n-vertex graph to maximize the number of edges incident with S. For almost all graphs with n vertices and all positive integers $k \leqslant n^\alpha$ where $\alpha < 1/6$, the k vertices of largest degree generate an optimal solution.

D. Hochbaum (1982). *Easy Solutions for the k-Center Problem or the Dominating Set Problem on Random Graphs,* School of Business Administration, University of California, Berkeley.

A set of vertices in a graph is *dominating* if each vertex not in the set is adjacent to some vertex in the set. Thus maximal stable sets are dominating. This paper considers the average behavior of a problem related to dominating sets.

J. Schmidt-Pruzan, E. Shamir, E. Upfal (1984). Random hypergraph coloring algorithms and the weak chromatic number. *J. Combin. Theory Ser. B.*
J. Schmidt-Pruzan (1983). *Probabilistic Analysis of Strong Hypergraph Coloring Algorithms and the Strong Chromatic Number,* Department of Applied Mathematics, Weizmann Institute of Science, Rehovot.

Algorithms are proposed for various hypergraph coloring problems. For certain random hypergraphs they are a.s. optimal to within a small constant factor.

R.M. MacGregor (1978). *On Partitioning a Graph: a Theoretical and Empirical Study,* Memorandum UCB/ERL M78/14, Electronics Research Laboratory, University of California, Berkeley.
T. Bui, S. Chaudhuri, T. Leighton, M. Sipser (1984). Graph bisection algorithms with good average case behavior. *Proc. 25th Annual IEEE Symp. Foundations of Computer Science.*

The *graph k-partition* problem involves the determination of a mimimum set of edges whose removal disconnects the graph into k equal-sized subgraphs. MacGregor investigates the performance of iterative improvement schemes and also provides probabilistic lower and upper bounds on the size of a minimum 2-partition. Bui *et al.* propose a polynomial-time algorithm and show that it finds a minimum 2-partition with high probability. They use a special sample space with the property that, with high probability, very few edges need to be deleted to partition the graph.

J.H. Reif, P.G. Spirakis (1980). Random matroids. *Proc. 12th Annual ACM Symp. Theory of Computing,* 385-397.

An analysis is made of the probability that a greedy algorithm will find a maximum independent set in a random independence system.

3.6. *Long paths*

It is easily seen that, given a constant $c>0$, the problem of determining if a graph with n vertices has a simple path of length at least cn is $\mathcal{NP}$-complete.

W.F. de la Vega (1979). Long paths in random graphs. *Studia Sci. Math. Hungar. 14,* 335-340.

A simple heuristic a.s. yields a path of length at least $(1-1.39/c)n$ in $G_{n,cn}$, and a similar results holds for random directed graphs.

M. Ajtai, J. Komlós, E. Szemerédi (1981). The longest path in a random graph. *Combinatorica 1,* 1-12.
B. Bollobás (1982). Long paths in sparse random graphs. *Combinatorica 2,* 223-228.

A.M. Frieze (1984B). *On Large Matchings and Cycles in Sparse Random Graphs,* Graduate School of Industrial Administration, Carnegie-Mellon University, Pittsburgh, PA.

These papers concern the existence of long paths rather than the analysis of algorithms. Frieze proves that $G_{n,c/n}$ a.s. contains a cycle of length $n(1-(1+\epsilon(c))ce^{-c})$, where $\epsilon(c) \rightarrow 0$ as $c \rightarrow \infty$.

3.7. *Hamiltonian cycles*

A Hamiltonian cycle in a graph or a directed graph is a closed path that passes through each vertex exactly once. It is a classical result that the problem of determining if a graph has a Hamiltonian cycle is $\mathcal{NP}$-complete.

V. Chvátal (1985). Hamiltonian cycles. E.L. Lawler, J.K. Lenstra, A.H.G. Rinnooy Kan, D.B. Shmoys (eds.). *The Traveling Salesman Problem,* Wiley, Chichester, Ch.11.

Section 3 of this chapter is an excellent survey of the research on the existence of Hamiltonian cycles in random graphs. It includes Karp's version and analysis of the *extension-rotation* algorithm. Other surveys are included in [Angluin & Valiant 1979] below and [Slominski 1982] (§1). See also [Johnson 1984] (§1).

L. Pósa (1976). Hamiltonian circuits in random graphs. *Discrete Math. 14,* 359-364.

There is a.s. a Hamiltonian cycle in the random graph $G_{n,N}$ with a number of edges $N \geqslant cn \log n$ for a suitable constant c. The proof is not based on an algorithm, but the extension-rotation idea can lead to good algorithms.

A.D. Koršunov (1976). Solution of a problem of Erdös and Rényi on Hamiltonian cycles in non-oriented graphs. *Soviet Math. Dokl. 17,* 760-764.

Consider the random graph $G_{n,N}$ with $N \geqslant n(\log n + \log\log n + \omega(n))/2$, where $\omega(n) \rightarrow \infty$ as $n \rightarrow \infty$. It is indicated that an algorithm a.s. finds a Hamiltonian cycle.

D. Angluin, L.G. Valiant (1979). Fast probabilistic algorithms for Hamiltonian circuits and matchings. *J. Comput. System Sci. 19,* 155-193.

Fast heuristics ($O(n \log^2 n)$ time) are shown to find Hamiltonian cycles a.s. in random graphs $G_{n,N}$ or similar random directed graphs when $N \geqslant cn \log n$ for a suitable constant c.

E. Shamir (1983). How many random edges make a graph Hamiltonian? *Combinatorica 3,* 123-131.

For the random graph $G_{n,p}$ with $p = p(n) = n^{-1}(\log n + c \log\log n)$, $c > 3$, there is an extension-rotation procedure which a.s. finds a Hamiltonian path within $O(n^2)$ steps.

The following five papers are concerned more with existence than with algorithms. They all use the extension-rotation idea.

J. Komlós, E. Szemerédi (1983). Limit distribution for the existence of hamiltonian cycles in random graphs. *Discrete Math. 43,* 55-63.

The random graph $G_{n,N}$ with N about $\frac{1}{2}n \log n + \frac{1}{2}n \log\log n + cn$ is Hamiltonian with probability tending to $\exp(\exp(-2c))$ as $n \to \infty$.

T.I. Fenner, A.M. Frieze (1983). On the existence of Hamiltonian cycles in a class of random graphs. *Discrete Math. 45,* 301-305.

The random graph G_n^m (see [Frieze 1984B], §3.3) is a.s. Hamiltonian for $m \geqslant 23$.

T.I. Fenner, A.M. Frieze (1982). *Hamiltonian Cycles in Random Regular Graphs,* Queen Mary College, University of London.
B. Bollobás (1983). Almost all regular graphs are Hamiltonian. *European J. Combin. 4,* 97-106.

In both these papers it is shown that for fixed k sufficiently large, the proportion of k-regular graphs on n vertices which are Hamiltonian tends to 1 as $n \to \infty$.

A.M. Frieze (1982). *Limit Distribution for the Existence of Hamiltonian Cycles in Random Bipartite Graphs,* Department of Computer Science and Statistics, Queen Mary College, University of London.

Let P_n be the probability that there is a Hamiltonian cycle in the random bipartite graph with $2n$ vertices in which the n^2 possible edges occur independently with probability $n^{-1}(\log n + \log\log n + c_n)$. Then $P_n \to 0$ if $c_n \to -\infty$, $P_n \to 1$ if $c_n \to \infty$ and $P_n \to \exp(-2e^{-c})$ if $c_n \to c$.

R.W. Robinson, N.C. Wormald (to appear). Almost all bipartite cubic graphs are Hamiltonian. *Proc. Silver Jubilee Conf. Combinatorics, Waterloo,* Academic Press, New York.

The proportion of bipartite labeled cubic graphs which are Hamiltonian tends to 1 as $n \to \infty$. The proof uses the second moment method.

3.8. *Bandwidth*

The *bandwidth* of a graph is the minimum over all labelings of the vertices with distinct integers of the maximum difference of the labels of adjacent vertices. The problem of determining the bandwidth is $\mathcal{NP}$-hard.

P.Z. Chinn, J. Chvátalovà, A.K. Dewdney, N.E. Gibbs (1982). The bandwidth problem for graphs and matrices - a survey. *J. Graph Theory 6,* 223-254.

There is a brief mention of the average-case complexity of bandwidth algorithms.

J. Turner (1983). Probabilistic analysis of bandwidth minimization algorithms. *Proc. 15th Annual ACM Symp. Theory of Computing*, 467-476.

This paper provides a probabilistic explanation of the effectiveness of *level algorithms* for bandwidth minimization on certain classes of graphs.

4. WEIGHTED GRAPHS

4.1. *Shortest paths*

The shortest paths problem is the problem of finding minimum weight paths between specified source vertices and destination vertices in a graph or digraph with weighted edges. In probabilistic analyses the edge weights are often taken to be independent identically distributed random variables.

P.M. Spira (1973). A new algorithm for finding all shortest paths in a graph of positive arcs in average time $O(n^2 \log^2 n)$. *SIAM J. Comput. 2,* 28-32.

A new algorithm is presented for the all pairs shortest path problem in a digraph with nonnegative edge weights. If the weights are drawn independently from a nonnegative continuous distribution, then the expected execution time is as stated in the title. This is one of the earliest papers to conduct a sound probabilistic analysis of an interesting combinatorial algorithm.

Y. Perl (1977). *Average Analysis of Simple Path Algorithms,* Technical report UIUCDCS-R-77-905, Department of Computer Science, University of Illinois at Urbana-Champaign.

In a random graph with n vertices and m edges, the expected number of edges inspected in searching for a path from a source vertex to a destination vertex using breadth-first or depth-first search is $O(n)$. In a random graph with edge weights drawn independently from a common nonnegative distribution, Prim's minimum spanning tree algorithm and Dijkstra's shortest path algorithm both run in expected time $O(n \log n \log(m/n))$ and Kruskal's minimum spanning tree algorithm runs in expected time $O(n \log n \log m)$.

P.A. Bloniarz, R.M. Meyer, M.J. Fischer (1979). *Some Observations on Spira's Shortest Path Algorithm,* Technical report 79-6, Computer Science Department, State University of New York, Albany.

Considers Spira's algorithm (see above) for the case that edges may have equal weights.

P. Bloniarz (1983). A shortest-path algorithm with expected time $O(n^2 \log n \log^* n)$. *SIAM J. Comput. 12,* 588-600.

Spira's algorithm is refined and the expected execution time is improved.

A.M. Frieze, G.R. Grimmett (1983). *The Shortest-Path Problem for Graphs with Random Arc Lengths,* School of Mathematics, University of Bristol.

For 'endpoint-independent' distributions, a modification of Spira's algorithm runs in $O(n(m+n \log n))$ expected time, where m is the expected number of edges with finite weight. When edge weights are independent with certain distributions, a further modification runs in $O(n^2 \log n)$ expected time.

M. Luby, P. Ragde (1983). *Bidirectional Search is $O(\sqrt{n})$ Faster Than Dijkstra's Shortest Path Algorithm,* Computer Science Division, University of California, Berkeley.

On a complete n-vertex digraph with independent exponentially distributed edge weights a variant of bidirectional search finds the shortest path from a given source to a given destination in expected time $O(n^{½} \log n)$. The algorithm also has a preprocessing phase requiring $O(n^2)$ expected time.

4.2. *Spanning trees*

We here consider the problem of finding a spanning tree of minimum weight in a graph with weighted edges. In probabilistic analyses the edge weights are usually independent identically distributed random variables.

Y. Perl (1977). *Average Analysis of Simple Path Algorithms,* Technical report UIUCDCS-R-77-905, Department of Computer Science, University of Illinois at Urbana-Champaign.

See §4.1.

R.M. Karp, R.E. Tarjan (1980). Linear expected time algorithms for connectivity problems. *J. Algorithms 1,* 374-393.

See §3.2 for a review of the first part of this paper. For random graphs with m edges and edge weights drawn independently from a common distribution, an algorithm is given which finds a minimum spanning forest in expected time $O(m)$.

A.M. Frieze (1982). *On the Value of a Random Minimum Spanning Tree Problem,* Technical report, Department of Computer Science and Statistics, Queen Mary College, University of London.

In a complete graph in which the weights of the edges are drawn independently from the uniform distribution on [0,1], the expected cost of the minimum spanning tree is asymptotic to $\Sigma_{k=1}^{\infty} 1/k^3 = 1.202...$

4.3. *Linear assignment*

An instance of the assignment problem is specified by an $n \times n$ real matrix (d_{ij}). An assignment is a permutation σ of $\{1,2, \ldots, n\}$. The cost of assignment σ is $\Sigma_{i=1}^{n} d_{i\sigma(i)}$. An optimal assignment is one of minimum cost.

The assignment problem can be viewed as the problem of finding a minimum weight perfect matching in a bipartite graph with n vertices in each part. The assignment problem is often used as a relaxation of the traveling salesman problem, in which σ is required to be a cyclic permutation. An optimal assignment can be found in $O(n^3)$ time using network flow techniques.

A.A. Borovkov (1962). Toward a probabilistic formulation of two problems from economy. *Math. Dokl. Akad. Nauk SSSR 146,* 983-986.

The d_{ij} are assumed to be drawn independently from a common distribution satisfying certain technical conditions. A greedy algorithm of complexity $O(n^2)$ has the property that, for every $\epsilon>0$, $Pr\{\text{GREEDY}\geqslant(1+\epsilon)\text{OPT}\}\rightarrow 0$ as $n\rightarrow\infty$. Here GREEDY is the cost of the solution produced by the greedy algorithm and OPT is the cost of an optimal assignment. This is one of the earliest publications concerned with the probabilistic analysis of approximation algorithms.

W.E. Donath (1969). Algorithm and average-value bounds for assignment problems. *IBM J. Res. Develop. 13,* 380-386.

An argument is offered suggesting that, when the d_{ij} are drawn independently from the uniform distribution over [0,1], the expected cost of an optimal assignment is less than 2.37. Certain conditioning effects are neglected. Computational results are presented indicating that the expected cost of an optimal assignment is close to 1.6 when n is large. Other probability distributions of the d_{ij} are also considered.

A.J. Lazarus (1979). *The Assignment Problem with Uniform* (0,1) *Cost Matrix,* B.A. thesis, Department of Mathematics, Princeton University, Princeton, NJ.

Let the d_{ij} be independent and uniformly distributed over [0,1]. Let Y_n be the expected cost of an optimal assignment and $Y = \lim_{n\rightarrow\infty} Y_n$. It is shown that $Y_2 = 23/30$ and $Y \geqslant 1+(1/e)$. Empirical evidence that $Y<\infty$ is given.

D.W. Walkup (1979). On the expected value of a random assignment problem. *SIAM J. Comput. 8,* 440-442.

If the d_{ij} are drawn independently from the uniform distribution over [0,1] then, with high probability, the cost of the optimal assignment is less than 3.

R.M. Karp (1980). An algorithm to solve the $m\times n$ assignment problem in expected time $O(mn \log n)$. *Networks 10,* 143-152.

An algorithm for the construction of an optimal $m\times n$ assignment is presented. If the elements of each row of (d_{ij}) are independent identically distributed random variables then the expected execution time of the algorithm is $O(mn \log n)$. The best guaranteed time bound known for an assignment algorithm is $O(m^2 n)$.

R. Loulou (1982). *Average Behavior of Heuristic and Optimal Solutions to the Maximization Assignment Problem,* Faculty of Management, McGill University, Montreal.

Let the d_{ij} be independent and exponentially distributed with rate λ. Let Z be the expected cost of a maximum-cost assignment. Then $1+1/n-1/\log n \leqslant \lambda Z/(n \log n) \leqslant 1+1/n+1/\log n$.

R.M. Karp (1984). *An Upper Bound on the Expected Cost of an Optimal Assignment,* Computer Science Division, University of California, Berkeley.

When the d_{ij} are drawn independently from the uniform distribution over [0,1], the expected cost of the optimal assignment is less than 2.

J.B.G. Frenk, A.H.G. Rinnooy Kan (1984). *Order Statistics and the Linear Assignment Problem,* Econometric Institute, Erasmus University, Rotterdam.

For a wide range of distribution functions F, the expected cost of the optimal assignment is asymptotic to $nF^{-1}(1/n)$.

4.4. *Network flow*

G.R. Grimmett, D.J.A. Welsh (1982). Flow in networks with random capacities. *Stochastics 7,* 205-229.
G.R. Grimmett, H.-C.S. Suen (1982). The maximal flow through a directed graph with random capacities. *Stochastics 8,* 153-159.
G.R. Grimmett, H. Kesten (1982). *First Passage Percolation, Network Flows and Electrical Resistances,* School of Mathematics, University of Bristol.

These papers prove existential results concerning the maximum value of a flow through certain randomly capacitated networks.

4.5. *Asymmetric traveling salesman*

An instance of the asymmetric traveling salesman problem is specified by an $n \times n$ matrix (d_{ij}). The object is to find a cyclic permutation σ to minimize $\Sigma_{i=1}^{n} d_{i\sigma(i)}$. If d_{ij} is interpreted as the distance from city i to city j then the problem amounts to finding a closed tour of minimum total distance which passes through each city exactly once. The problem is $\mathcal{NP}$-hard, and it is known that a polynomial-time approximation algorithm with good worst case performance does not exist unless $\mathcal{P} = \mathcal{NP}$. Probabilistic analyses of the problem usually assume that the d_{ij} are drawn independently from a common distribution. Another variant, considered in §5.5, is the Euclidean traveling salesman problem, in which the cities are points in the plane and distance is Euclidean distance.

M. Bellmore, J.C. Malone (1971). Pathology of traveling-salesman subtour-elimination algorithms. *Oper. Res. 19,* 278-307, 1766.

A branch-and-bound algorithm based on subtour elimination is claimed to solve random asymmetric traveling salesman problems in a polynomial-bounded expected number of steps. The argument neglects certain conditioning effects.

E.H. Guimady, V.A. Pereplitsa (1974). An asymptotical approach to solving the traveling salesman problem. *Upravljaemye Sistemy,* 35-45.

The nearest neighbor algorithm is analyzed on the asssumption that the d_{ij} are drawn independently from a uniform distribution.

J.K. Lenstra, A.H.G. Rinnooy Kan (1978). On the expected performance of branch-and-bound algorithms. *Oper. Res. 26,* 347-349.

A flaw is pointed out in the alleged proof of [Bellmore & Malone 1971] that a branch-and-bound algorithm for the asymmetric traveling salesman problem runs in polynomial expected time.

T. Lcipala (1978). On the solutions of stochastic traveling salesman problems. *European J. Oper. Res. 2,* 291-297.

Stochastic upper and lower bounds on the length of the optimal tour are derived under various probabilistic assumptions. The upper bounds come from an analysis of the nearest neighbor algorithm.

R.S. Garfinkel, K.C. Gilbert (1978). The bottleneck traveling salesman problem: algorithms and probabilistic analysis. *J. Assoc. Comput. Mach. 25,* 435-448.

The bottleneck traveling salesman problem asks for a tour in which the weight of the heaviest arc is minimized. The distribution of the cost of such a tour is studied under the assumption that the edge weights are drawn independently from a common distribution. The results are closely related to recent work on the existence of Hamiltonian cycles in random digraphs (see §3.7).

R.M. Karp (1979). A patching algorithm for the nonsymmetric traveling-salesman problem. *SIAM J. Comput 8,* 561-573.

A patching algorithm is given which solves the asymmetric traveling salesman problem by solving the assignment problem and then patching the cycles of the optimal assignment together to form a tour. If the d_{ij} are drawn independently from the uniform distribution over [0,1], then the expected ratio between the cost of the tour produced and the cost of the optimal tour tends to 1. This result is refined in [Karp & Steele 1985] (see below).

J.-C. Panayiotopoulos (1982). Probabilistic analysis of solving the assignment problem for the traveling salesman problem. *European J. Oper. Res. 9,* 77-82.

The traveling salesman problem is solved by generating the assignments in order of increasing costs, until the first cyclic permutation is found. The analysis of this method ignores certain conditioning effects.

R.M. Karp, J.M. Steele (1985). Probabilistic analysis of heuristics. E.L. Lawler, J.K. Lenstra, A.H.G. Rinnooy Kan, D.B. Shmoys (eds.). *The Traveling Salesman Problem,* Wiley, Chichester, Ch. 6.

This chapter surveys and extends the existing results on the probabilistic analysis of the Euclidean and the asymmetric traveling salesman problems. For the first problem, see §5.5. For the second, a variant of the patching algorithm from [Karp 1979] is analyzed. It is shown that, if the d_{ij} are drawn independently from the uniform distribution over [0,1], then E[(PATCH−OPT)/OPT] $= O(n^{-1/2})$. Here OPT is the cost of an optimal tour and PATCH is the cost of the tour produced by the patching algorithm.

4.6. *Quadratic assignment*

The quadratic assignment problem generalizes many $\mathcal{NP}$-hard combinatorial optimization problems, including the traveling salesman problem. An instance is specified by two $n \times n$ matrices (a_{ij}) and (b_{ij}). In the standard quadratic assignment problem the objective is to find a permutation σ that minimizes $\Sigma_{i,j} a_{ij} b_{\sigma(i)\sigma(j)}$. In the *bottleneck* quadratic assignment problem the objective is to minimize over all σ $\max_{i,j} \{a_{ij} b_{\sigma(i)\sigma(j)}\}$. The problem is called *planar* if the a_{ij} are distances between points in the plane.

R.E. Burkard, U. Fincke (1982A). On random quadratic bottleneck assignment problems. *Math. Programming 23,* 227-232.

R.E. Burkard, U. Fincke (1982B). Probabilistic asymptotic properties of quadratic assignment problems. *Z. Oper. Res. 27,* 73-81.

The ratio between the maximum and the minimum cost of an assignment a.s. tends to 1 as $n \to \infty$. This holds for the standard as well as the bottleneck problem, both when the a_{ij} are arbitrary and when they are planar distances (for any L_q-norm).

J.B.G. Frenk, M. van Houweninge, A.H.G. Rinnooy Kan (1982). *Asymptotic Properties of Assignment Problems,* Econometric Institute, Erasmus University, Rotterdam.

The results from [Burkard & Fincke 1982B] are generalized. E.g., an explicit expression for the asymptotic optimal solution value is obtained.

4.7. *Miscellaneous*

B.W. Weide (1980). Random graphs and graph optimization problems. *SIAM J. Comput. 9,* 552-557.

A method is presented to relate results regarding the probability of existence of certain subgraphs in random graphs $G_{n,p}$ to the probabilistic behavior of the optimal solution value to graph optimization problems, where the edge weights are independently chosen from an arbitrary distribution. The method is illustrated on the bottleneck and the standard traveling salesman problem.

G.S. Lueker (1981). Optimization problems on graphs with independent random edge weights. *SIAM J. Comput. 10,* 338-351.

Optimization problems are considered in which the input is a complete graph on n vertices together with a numerical weight for each edge. The edge weights are assumed to be drawn independently from the normal distribution with mean 0 and variance 1. The expected value of the maximum cost of a traveling salesman tour and the expected value of the cost of a maximum spanning tree are both asymptotic to $n\sqrt{(2 \log n)}$. If k is fixed then the expected maximum cost of a k-clique is asymptotic to $k\sqrt{((k-1)\log n)}$. For the traveling salesman problem, a simple greedy algorithm achieves the same expected asymptotic behavior as the optimal solution. A greedy algorithm for the maximum weight clique problem is also considered.

R.E. Burkard, U. Fincke (1982). *Probabilistic Asymptotic Properties of Some Combinatorial Optimization Problems,* Bericht 82-3, Institut für Mathematik, Technische Universität Graz.

Continuing their work reviewed in §4.6, the authors consider problems of the form $\min_{S \in T}\{\Sigma_{e \in S} c(e)\}$ and $\min_{S \in T}\{\max_{e \in S}\{c(e)\}\}$, where T is a family of subsets of an m-element set and each element e in the universe has a cost $c(e)$. The costs are assumed to be drawn independently from the uniform distribution over [0,1]. Conditions are given under which, for every $\epsilon>0$, the probability tends to 1 as $m \rightarrow \infty$ that the ratio between the cost of the worst solution and the cost of the best solution is less than $1+\epsilon$. This result applies to quadratic assignment problems as well as to certain network flow, linear programming and matching problems.

5. Euclidean Problems

This section is concerned with combinatorial optimization problems whose specification includes a set of points in Euclidean space. Probabilistic analyses of optimal solution values and approximation algorithms for such problems often start from the assumption that the points are independent and uniformly distributed over a fixed 2-dimensional region, e.g. a circle or a square. Many results can be extended to other distributions and to higher dimensions.

In addition to the Euclidean problems dealt with below, results have been obtained for Euclidean *quadratic assignment* problems. These have been reviewed in §4.6.

5.1. *Closest points*

J.L. Bentley, M.I. Shamos (1978). Divide and conquer for linear expected time. *Inform. Process. Lett. 7,* 87-91.

A divide-and-conquer scheme is proposed to find the convex hull of n points in the plane in $O(n \log n)$ worst case and $O(n)$ expected time.

J.L. Bentley, B.W. Weide, A.C. Yao (1980). Optimal expected-time algorithms for closest-point problems. *ACM Trans. Math. Software 6,* 563-580.

A basic approach to solve closest point problems is the *cell technique,* which partitions the region containing the n points into cells with a constant expected number of points per cell. Spiral search of neighboring cells solves the nearest neighbor problem for arbitrary dimension in $O(1)$ expected time. In the 2-dimensional case, the cell technique allows construction of the Voronoi diagram, and hence of the minimum spanning tree, in $O(n)$ expected time. In higher dimensions, the minimum spanning tree problem requires $O(n \log\log n)$ time, and it is an open question if $O(n)$ expected time can be achieved.

P. Lehert (1981). Clustering by connected components in $O(n)$ expected time. *RAIRO Inform. 15,* 207-218.

Connected components of a graph, defined by a threshold distance, are found through the cell technique in linear expected time, provided that the L_∞-metric is used.

J.M. Steele (1982). Optimal triangulation of random samples in the plane. *Ann. Probab. 10,* 548-553.

The length of a minimal triangulation of n points in the Euclidean plane is a.s. asymptotic to $\alpha\sqrt{n}$ for some constant α. This settles a conjecture of G. Turán.

5.2. *Shortest paths*

R. Sedgewick, J.S. Vitter (1984). Shortest paths in Euclidean graphs. *Proc. 25th Annual IEEE Symp. Foundations of Computer Science.*

For a variety of Euclidean random graph models with n vertices and m edges, the authors' algorithm finds the shortest path between a specified pair of vertices in $O(n)$ expected time. Classical algorithms require $O(n^2)$ time for dense graphs and $O(n \log^2 n)$ time for sparse graphs on average.

5.3. *Matching*

C.H. Papadimitriou (1978). The probabilistic analysis of matching heuristics. *Proc. 15th Annual Allerton Conf. Communication, Control and Computing,* 368-378.

Steele's asymptotic result for subadditive Euclidean functionals (see [Steele 1981B], §5.5) implies that the optimal value of a Euclidean matching is a.s. asymptotic to $\beta\sqrt{n}$, for some constant β. It is shown that $0.25 \leqslant \beta \leqslant 0.40106$ and conjectured that $\beta \simeq 0.35$.

5.4. *Location*

Two basic location problems are the *K-median* and the *K-center* problem, in which K locations from among n given points are to be chosen so as to minimize the sum and the maximum respectively of the distances of each point to the nearest location. Asymptotic analyses for these problems necessarily involve an assumption about the growth rate of K as a function of n.

M.L. Fisher, D.S. Hochbaum (1980). Probabilistic analysis of the planar K-median problem. *Math. Oper. Res. 5,* 27-34.

A *partitioning* heuristic, much in the spirit of Karp's traveling salesman heuristic (see [Karp 1977], §5.5), splits the square into congruent subsquares and solves the weighted K-median problem on these subsquares, with the weight of each subsquare equal to the number of points that it contains. The heuristic is asymptotically optimal in probability, and the asymptotic optimal solution value lies a.s. in $[\gamma' n / \sqrt{K}, \gamma'' n / \sqrt{K}]$, for constants γ', γ''. (Cf. Ch.9, §5.2.)

C.H. Papadimitriou (1981). Worst-case and probabilistic analysis of a geometric location problem. *SIAM J. Comput. 10,* 542-557.

For the case that $K = o(n / \log n)$, a *honeycomb* heuristic for the K-median problem, which divides the region into regular hexagons, is asymptotically optimal in probability, and the optimal solution value is a.s. asymptotic to $\gamma n / \sqrt{K}$, with $\gamma = 2^{1/4}/3^{3/4}$. The proof is based on the observation that, as $n \to \infty$, the discrete problem approaches the continuous problem in which demand is not concentrated at points but spread uniformly over the region. For the case that $K = \Theta(n)$, a *partitioning* heuristic is asymptotically optimal in probability, under the assumption that the points are generated by a Poisson process. (Cf. Ch.9, §5.2.)

D.S. Hochbaum, J.M. Steele (1981). Steinhaus' geometric location problem for random samples in the plane. *Adv. in Appl. Probab. 14,* 56-67.

For the case that $K = \Theta(n)$, Steele's asymptotic result for subadditive Euclidean functionals (see [Steele 1981B], §5.5) is extended to the K-median problem, so that the optimal solution value is a.s. asymptotic to $\delta \sqrt{n}$, for some constant δ. We note that for $K = \Theta(n)$ the K-center problem is still open and that for neither model an a.s. asymptotically optimal heuristic is known.

E. Zemel (1984). Probabilistic analysis of geometric location problems. *Ann. Oper. Res. 1.*

For the case that $K = o(n / \log n)$, the *honeycomb* heuristic is a.s. asymptotically optimal for both the K-median and the K-center problem; the relation to the continuous version of these problems again provides the clue to the analysis. The optimal solution value for the K-center problem is a.s. asymptotic to $\epsilon / \sqrt{K}$, with $\epsilon \simeq 0.377$.

We conclude this subsection with a paper on a related problem.

C.H. Papadimitriou (1978). The complexity of the capacitated tree problem. *Networks 8,* 217-230.

The capacitated tree problem is to link customers to a depot by means of a spanning tree such that deletion of the depot yields components of bounded cardinality. The proof that the proposed partitioning heuristic is a.s. asymptotically optimal neglects the dependence of the region over which the asymptotic analysis is carried out on the actual sample.

5.5. *Routing*

The seminal work in the probabilistic analysis of Euclidean problems has been carried out in the context of routing problems and, more specifically, of the Euclidean traveling salesman problem.

J. Beardwood, J.H. Halton, J.M. Hammersley (1959). The shortest path through many points. *Proc. Cambridge Philos. Soc. 55,* 299-327.

The length of the optimal traveling salesman tour through n cities is a.s. asymptotic to $\zeta\sqrt{n}$, where ζ depends on the size and shape of the region and on the probability distribution of the cities over it.

R.M. Karp (1977). Probabilistic analysis of partitioning algorithms for the traveling-salesman problem in the plane. *Math. Oper. Res. 2,* 209-224.

Karp's *partitioning* heuristic constructs a tour by dividing the region into subregions, each containing a small number of points, and then linking optimal tours through each subregion together. The absolute error of the heuristic grows more slowly than $\sqrt{n}$. Hence, the above result from Beardwood *et al.* implies that the relative error tends to 0 a.s. Depending on whether or not the partitioning scheme takes the actual location of the cities into account, the running time of the method is in expectation or deterministically polynomial.

J.M. Steele (1981A). Complete convergence of short paths and Karp's algorithm for the TSP. *Math. Oper. Res. 6,* 374-378.

Steele establishes complete convergence for the result of Beardwood *c.s.*, something that Karp in his above paper had tacitly assumed.

J.M. Steele (1981B). Subadditive Euclidean functionals and nonlinear growth in geometric probability. *Ann. Probab. 9,* 365-376.

A generalization of the BHH result is established for a class of subadditive functions defined on random sets of independently distributed points. Examples of such functions are the lengths of an optimal traveling salesman tour, of a rectilinear Steiner tree, and of a minimum matching.

J.H. Halton, R. Terada (1982). A fast algorithm for the Euclidean traveling salesman problem, optimal with probability one. *SIAM J. Comput. 11,* 28-46.

A partitioning method, similar to Karp's approach, yields asymptotically optimal tours a.s., whereas its running time is almost linear in probability.

R.M. Karp, J.M. Steele (1985). Probabilistic analysis of heuristics. E.L. Lawler, J.K. Lenstra, A.H.G. Rinnooy Kan, D.B. Shmoys (eds.). *The Traveling Salesman Problem,* Wiley, Chichester, Ch. 6.

This chapter reviews probabilistic analyses for traveling salesman problems. See also §4.5.

J.M. Steele (to appear). Probabilistic algorithm for the directed traveling salesman problem. *Math. Oper. Res.*

In a model for random Euclidean asymmetric traveling salesman problems, the expected optimal tour length is shown to be asymptotic to $\eta\sqrt{n}$. A partitioning heuristic is presented whose relative error tends to 0 in expectation.

A. Marchetti-Spaccamela, A.H.G. Rinnooy Kan, L. Stougie (to appear). Hierarchical routing problems. *Networks.*

The two-stage planning problem under consideration asks for the determination of a number of vehicles minimizing the sum of the acquisition costs and the length of the longest tour through the customers assigned to any vehicle. At the time of acquiring the vehicles, all that is known about the customers is that they are uniformly distributed over a circular region. The asymptotic optimal solution value and an asymptotically optimal heuristic are derived. See Ch.11, §12.2 for related material.

M. Haimovich, A.H.G. Rinnooy Kan (to appear). Bounds and heuristics for capacitated routing problems. *Math. Oper. Res.*

A probabilistic value analysis for a capacitated Euclidean multisalesmen problem is presented, together with several heuristics whose relative error tends to 0 a.s.

6. Linear Programming

The linear programming problem requires the minimization of a linear function subject to linear constraints. It can be written in the form $\min_x \{c^T x : Ax \geqslant b, x \geqslant 0\}$, where $c, x \in \mathbb{R}^d$, $A \in \mathbb{R}^{m \times d}$, and $b \in \mathbb{R}^m$.

The outstanding practical experience obtained with the *simplex method* for linear programming is in sharp contrast to its exponential worst case behavior. Only recently, a sequence of papers has appeared that provides a beginning of an analytical explanation of the method's efficiency, by demonstrating that the expected number of pivots for certain simplex variants is bounded by a polynomial function of d and m.

K.H. Borgwardt (1982). Some distribution-independent results about the asymptotic order of the average number of pivot steps of the simplex method.

Math. Oper. Res. 7, 441-462.
K.H. Borgwardt (1982). The average number of pivot steps required by the simplex-method is polynomial. *Z. Oper. Res. 26,* 157-177.

Borgwardt investigates linear programs of the form $\max\{c^T x : Ax \leqslant e\}$, where $c, x \in \mathbb{R}^d$ and $A \in \mathbb{R}^{n \times d}$ $(n \geqslant d)$. Let A_i^T denote the ith row of A. The behavior of a parametric simplex variant (the *shadow vertex* algorithm) is analyzed in a probabilistic model in which the vectors $c, A_1, \ldots, A_n$ are independently drawn from the same spherically symmetric distribution. All problems generated in this way are feasible. In the second paper, the expected number of pivots is shown to be $O(d^4 n)$.

These papers received the 1982 Lanchester prize of the Operations Research Society of America.

S. Smale (1983). On the average speed of the simplex method of linear programming. *Math. Programming 27,* 241-262.
S. Smale (1983). The problem of the average speed of the simplex method. A. Bachem, M. Grötschel, B. Korte (eds.). *Mathematical Programming: the State of the Art - Bonn 1982,* Springer, Berlin, 530-539.

Smale analyzes the behavior of Dantzig's *self-dual* simplex method, when viewed as a special case of Lemke's algorithm for the linear complementarity problem, in a probabilistic model in which the data are independently drawn from a spherically symmetric distribution. This model does not allow any kind of degeneracy. The expected number of pivots is shown to be $O(c_d (\log m)^{d(d+1)})$, where c_d depends superexponentially on d.

C. Blair (1983). *Random Linear Programs with Many Variables and Few Constraints,* Working paper 946, College of Business Administration, University of Illinois, Urbana-Champaign.

A much simplified proof for a slightly weaker version of Smale's results.

M. Haimovich (1983). *The Simplex Method is Very Good! - On the Expected Number of Pivot Steps and Related Properties of Random Linear Programs,* Columbia University, New York.
I. Adler (1983). *The Expected Number of Pivots Needed to Solve Parametric Linear Programs and the Efficiency of the Self-Dual Simplex Method,* Department of Industrial Engineering and Operations Research, University of California, Berkeley.

Adler and Haimovich independently investigate the length of a path generated by some parametric simplex variants, in a probabilistic model that requires only almost sure nondegeneracy and 'sign invariance', i.e., invariance of the distribution under changing the sign of any subset of rows or columns. This model enables a simpler probabilistic analysis in terms of elegant combinatorial counting arguments. The expected length of a path from worst to best solution is shown to be $\min\{m, d\} + 1$; but the analysis presupposes that a feasible initial vertex is given.

I. Adler, R.M. Karp, R. Shamir (1983). *A Family of Simplex Variants Solving an $m \times d$ Linear Program in Expected Number of Pivots Depending on d Only,* Report UCB CSD 83/157, Computer Science Division, University of California, Berkeley.

Adler, Karp & Shamir analyze a family of algorithms that proceed according to a *constraint-by-constraint* principle (and that include a whole family of known simplex variants) in a general probabilistic model that does not require sign invariance. The expected number of pivots is shown to be bounded by an exponential function of d, independent of m.

I. Adler, R.M. Karp, R. Shamir (1983). *A Simplex Variant Solving an $m \times d$ Linear Program in $O(\min(m^2,d^2))$ Expected Number of Pivot Steps,* Report UCB CSD 83/158, Computer Science Division, University of California, Berkeley.

I. Adler, N. Megiddo (1983). *A Simplex Algorithm Whose Average Number of Steps is Bounded Between Two Quadratic Functions of the Smaller Dimension.*

M.J. Todd (1983). *Polynomial Expected Behavior of a Pivoting Algorithm for Linear Complementarity and Linear Programming Problems,* Technical report 595, School of Operations Research and Industrial Engineering, Cornell University, Ithaca, NY.

Adler, Karp & Shamir analyze a parametric version of the *constraint-by-constraint* method and Adler & Megiddo and Todd analyze the *self-dual* algorithm in a probabilistic model that is closely related to that used by Adler and Haimovich. As Megiddo pointed out, the lexicographic versions that are subjected to these independent analyses execute the same sequence of pivots. The expected number of pivots is shown to be $O(\min\{m^2,d^2\})$. Under stronger probabilistic assumptions, Adler & Megiddo also obtain a quadratic *lower* bound.

R. Shamir (1984). *The Efficiency of the Simplex Method: a Survey,* Department of Industrial Engineering and Operations Research, University of California, Berkeley.

§6 of this survey presents a more detailed review and a useful assessment of the above material.

7. Packing and Covering

7.1. *Satisfiability and tiling*

The satisfiability problem is the problem of deciding whether there exists an assignment of truth values to variables that makes a given Boolean formula true and was proved to be $\mathcal{NP}$-complete by S.A. Cook in 1971. The tiling problem was proved to be $\mathcal{NP}$-complete by L.A. Levin, the second discoverer of $\mathcal{NP}$-completeness theory, in 1973. For more information on the material reviewed in this subsection and, in particular, for an exposition of Levin's results on the 'random tiling' problem reported in [Levin 1984], see [Johnson

1984] (§1).

J.M. Plotkin, J.W. Rosenthal (1978). On the expected number of branches in analytic tableaux analysis in propositional calculus. *Notices Amer. Math. Soc. 25,* A-437.

Let E_n denote the expected number of branches generated when the method is applied to a random formula in which AND, OR and NOT are the only connectives, negation is applied only to atomic formulas and there are n occurrences of AND and OR. Then $c(1.08)^n \leq E_n \leq d(1.125)^n$, where c and d are constants.

A. Goldberg, P. Purdom, C. Brown (1982). Average time analysis of simplified Davis-Putnam procedures. *Inform. Process. Lett. 15,* 72-75. Corrigendum. *Inform. Process. Lett. 16,* 213.

A random conjunctive normal form formula consists of n independently generated random clauses. In each clause, each of the m variables occurs positively with probability p, negatively with probability p, and is absent with probability $1-2p$. The expected time for a backtracking algorithm due to Davis & Putnam to determine whether a random formula is satisfiable is polynomial in n and m (but exponential in $1/p$).

J. Franco, M. Paull (1983). Probabilistic analysis of the Davis Putnam procedure for solving the satisfiability problem. *Discrete Appl. Math. 5,* 77-87.

For random formulas with n clauses, m variables and constant clause lengths, the expected running time of the Davis-Putnam procedure is exponential. Note that the above result of Goldberg *et al.* applies to a model in which the expected clause length is proportional to m.

L.A. Levin (1984). Problems, complete in 'average' instance. *Proc. 16th Annual ACM Symp. Theory of Computing,* 465.

Levin develops a theory of $\mathcal{NP}$-completeness for the average rather than the worst case. He defines a class RANDOM $\mathcal{NP}$ of pairs (X,μ), where X may be any problem in $\mathcal{NP}$ and μ is a probability measure whose distribution function is computable in polynomial time, and a new notion of polynomial transformation within this class, which transforms distributions as well as instances. He proves that RANDOM TILING, i.e., the tiling problem with a very natural distribution of its instances, is complete in RANDOM $\mathcal{NP}$.

7.2. *Bin packing*

E.G. Coffman, Jr., M.R. Garey, D.S. Johnson (1983). *Approximation Algorithms for Bin-Packing - An Updated Survey,* Bell Laboratories, Murray Hill, NJ.

A thorough survey with excellent coverage of recent deterministic and probabilistic results.

(a) 1-*dimensional bin packing*

This is the problem of packing n items into a minimum number of bins of capacity 1. Associated with each item is a positive real number less than 1 called its *size.* The sum of the sizes of the items packed into a single bin may not exceed the bin capacity. Associated with any packing is the wasted space, defined as the number of bins used minus the sum of the sizes of all the items.

In probabilistic analyses of this problem it is usually assumed that the item sizes are drawn independently from a common distribution. Often, the uniform distribution over [0,1], or, less commonly, over [a,b], is postulated. A probability distribution is said to allow *perfect packing* if, with probability tending to 1 as $n \to \infty$, the waste is bounded above by some function which is $o(n)$. The two algorithms most commonly studied are *next fit,* an on-line algorithm which packs the items in their given order and starts a new bin whenever the next item cannot fit in the present bin, and *first fit decreasing,* which considers the items in decreasing order of size and places each item in the first bin that can accept it. In addition to the probabilistic studies reported here there is a very extensive literature on the worst case performance of bin packing algorithms (see [Coffman, Garey & Johnson 1983] above).

S.D. Shapiro (1977). Performance of heuristic bin packing algorithms with segments of random length. *Inform. and Control 35,* 146-158.

An approximate analysis of the next fit algorithm is given, on the assumption that the item sizes are drawn independently from an exponential distribution.

E.G. Coffman, Jr., M. Hofri, K. So, A.C. Yao (1980). A stochastic model of bin packing. *Inform. and Control 44,* 105-115.

On the assumption that the item sizes are uniformly distributed over [0,1], an upper bound is derived on the expected number of bins required by the next fit algorithm.

G.N. Frederickson (1980). Probabilistic analysis for simple one- and two-dimensional bin packing algorithms. *Inform. Process. Lett. 11,* 156-161.

For a simple 1-dimensional bin-packing algorithm it is shown that the expected waste is $O(n^{2/3})$, on the assumption that the item sizes are uniformly distributed over [0,1]. Since the given algorithm always uses at least as many bins as the first fit decreasing or best fit heuristics, the expected waste for these heuristics is also $O(n^{2/3})$. Several authors later improved this bound to $O(n^{1/2})$. Implications for 2-dimensional strip packing are also explored.

W. Knödel (1981). A bin packing algorithm with complexity $O(n \log n)$ and performance 1 in the stochastic limit. J. Gruska, M. Chytil (eds.). *Mathematical Foundations of Computer Science 1981,* Lecture Notes in Computer Science 118, Springer, Berlin, 369-378.

A simple packing algorithm has expected waste $O(n^{½})$ when the item sizes are drawn independently from a decreasing probability distribution over [0,1].

R. Loulou (1982). *Probabilistic Behavior of Optimal Bin Packing Solutions,* Faculty of Management, McGill University, Montreal.

Every convex distribution function on [0,1] with bounded second derivative on [0,½] allows perfect packing.

M. Hofri (1982). *Bin-Packing: an Analysis of the Next-Fit Algorithm,* Technical report 242, Department of Computer Science, Technion, Haifa.

The next fit algorithm is analyzed on the assumption that the item sizes are uniformly distributed over [0,1]. Detailed information about the distribution of the number of bins is derived.

N. Karmarkar (1982). Probabilistic analysis of some bin-packing algorithms. *Proc. 23rd Annual IEEE Symp. Foundations of Computer Science,* 107-111.

On the assumption that the item sizes are drawn from the uniform distribution over $[0,a]$, a closed form expression is derived for the expected number of bins required by the next fit algorithm. It is also proved that the uniform distribution over $[0,a]$ permits perfect packing.

G.S. Lueker (1983). *An Average-Case Analysis of Bin Packing with Uniformly Distributed Item Sizes,* Report 181, Department of Information and Computer Science, University of California, Irvine.

When the item sizes are uniformly distributed over [0,1], a simple algorithm which never places more than two items in a bin achieves expected waste $O(n^{½})$.

G.S. Lueker (1983). Bin packing with items uniformly distributed over intervals $[a,b]$. *Proc. 24th Annual IEEE Symp. Foundations of Computer Science,* 289-297.

For a large class of choices of the interval $[a,b]$ it is determined whether the uniform distribution over $[a,b]$ allows perfect packing. The analysis makes interesting use of a linear programming technique.

J.L. Bentley, D.S. Johnson, F.T. Leighton, C.C. McGeoch, L.A. McGeoch (1984). Some unexpected expected behavior results for bin packing. *Proc 16th Annual Symp. Theory of Computing,* 279-288.

For the uniform distribution over [0,1], the first fit algorithm has expected waste $O(n^{0.8})$. For the uniform distribution over $[0,a]$, $a \leqslant ½$, the first fit decreasing algorithm has expected waste $O(1)$. These results were entirely unexpected and are of extraordinary interest.

P.W. Shor (1984). The average-case analysis of some on-line algorithms for bin packing. *Proc. 25th Annual IEEE Symp. Foundations of Computer Science.*

For the uniform distribution over [0,1], tighter upper and lower bounds on the performance of on-line algorithms are derived. The analysis is based on the relation between the bin packing problem and a 2-dimensional matching problem.

(b) *d-dimensional bin packing*

This is the problem of packing d-dimensional items into d-dimensional bins. The variants of the problem include rectangle packing, strip packing and vector packing. In *rectangle packing* the items are d-dimensional rectangles with sides parallel to the coordinate axes and each bin is a unit hypercube. In 2-dimensional *strip packing* the items are again rectangles with sides parallel to the coordinate axes but there is a single bin of fixed width and infinite height. The object is to minimize the vertical extent of the area used for packing. In *vector packing*, each bin satisfies the following constraint: for $i = 1,2,\ldots,d$, the sum of the ith coordinates of the items in the bin is less than or equal to 1.

M. Hofri (1980). Two-dimensional packing: expected performance of simple level algorithms. *Inform. and Control 45*, 1-17.

The problem of packing rectangles into a semi-infinite strip is discussed. The vertical and horizontal dimensions of the rectangles are assumed to be independent random variables drawn from the uniform distribution over [0,1]. The expected efficiencies of the next fit, rotatable next fit and next fit decreasing algorithms are derived.

R.M. Karp, M. Luby, A. Marchetti-Spaccamela (1984). A probabilistic analysis of multidimensional bin packing problems. *Proc. 16th Annual ACM Symp. Theory of Computing*, 289-298.

A simple algorithm for rectangle packing is analyzed on the assumption that the dimensions of the items are independent and uniformly distributed over [0,1]. In the case $d = 2$, the expected waste is $\Omega(\sqrt{(n \log n)})$ and $O(\sqrt{n} \log n)$. For $d \geqslant 3$, the expected waste is $\Theta(n^{(d-1)/d})$. Probabilistic analyses of algorithms for strip packing and vector packing are also given.

7.3. *Multiprocessor scheduling*

This is the problem of scheduling n tasks with known execution times on m identical parallel processors to minimize the makespan, defined as the maximum, over all processors, of the sum of the execution times of the tasks assigned to that processor. Probabilistic analyses usually assume that the execution times are independent identically distributed random variables.

The problem can be viewed as a variant of the 1-dimensional bin packing problem in which the number of bins is fixed and the capacity of the largest bin is to be minimized.

An important class of on-line multiprocessor scheduling algorithms is formed by the *list scheduling* algorithms, in which the tasks are arranged in a linear list, and, whenever a processor completes a task, the first unscheduled task on the list is assigned to that processor. The LPT algorithm, in which the list is arranged in decreasing order of execution times, is especially popular. Another class of algorithms based on a *differencing operation* has recently been found to have excellent properties (see [Karmarkar & Karp] below).

In addition to the probabilistic work surveyed here, there is an extensive literature on the worst case analysis of multiprocessor scheduling algorithms; see [Coffman, Garey & Johnson 1983] in §7.2 and Ch.11, §4.2.

For the probabilistic analysis of a *single-machine scheduling* algorithm, see [Gazmuri 1981] in Ch.11, §11.1. For the probabilistic analysis of *hierarchical scheduling* systems, in which a processor acquisition phase precedes the actual scheduling phase, see Ch.11, §12.2.

J.L. Bruno, P.J. Downey (1982). *Probabilistic Bounds on the Performance of List Scheduling,* Technical report TR 82-19, Computer Science Department, University of Arizona, Tucson.

If the execution times are drawn independently from a uniform distribution then, with probability $\geq 1-\epsilon$, the ratio of the makespan of an LPT schedule to the optimal makespan is less than $1+f(m,n,\epsilon)$, where $f(m,n,\epsilon) \simeq 1+(2(m-1)/n)$. The analysis enables concrete numerical results on the distribution of the relative error to be obtained for small values of n.

E.G. Coffman, Jr., E.N. Gilbert (1983). *On the Expected Relative Performance of List Scheduling,* Technical memorandum, Bell Laboratories, Murray Hill, NJ.

The analysis in the above paper is refined and extended to exponential distributions.

J.B.G. Frenk, A.H.G. Rinnooy Kan (1983). *The Asymptotic Optimality of the LPT Heuristic,* Econometric Institute, Erasmus University, Rotterdam.

Let OPT denote the cost of an optimal solution, and let LPT be the cost of the solution obtained by the longest processing time first heuristic. If the execution time distribution has a finite second moment and positive density at zero, then LPT-OPT converges a.s. and in expectation to 0 as $n \to \infty$.

R. Loulou (1984). Tight bounds and probabilistic analysis of two heuristics for parallel processor scheduling. *Math. Oper. Res. 9,* 142-150.

Let the execution times of the tasks be independent and identically distributed with finite mean. Let RLP be the cost of the solution obtained by applying list scheduling to a randomly ordered list of the tasks. Let m be the number of processors. When $m \geq 2$ the random variable RLP-OPT is stochastically bounded by a finite random variable for any value of n. When $m = 2$, a certain upper bound on this random variable converges in distribution as $n \to \infty$. When $m \geq 2$, LPT-OPT is stochastically smaller than a fixed random variable which does not depend on n.

N. Karmarkar, R.M. Karp (to appear). The differencing method of set partitioning. *Math. Oper. Res.*

A simple differencing algorithm is presented. If the item sizes are drawn independently from a smooth distribution with bounded support then, with probability tending to 1, the difference between the completion times of the last and first machines to complete is bounded above by a quantity of the form $n^{-c \log n}$.

N. Karmarkar, R.M. Karp, G.S. Lueker, A. Odlyzko (1984). *The Probability Distribution of the Optimal Value of a Partitioning Problem*, Bell Laboratories, Murray Hill, NJ.

The 2-processor scheduling problem is considered under the assumption that the execution times are drawn independently from the uniform distribution over [0,1]. Let D denote the minimum, over all partitions of the tasks into two sets, of the absolute difference between the sums of the execution times of the tasks in the two sets. Then, with probability tending to 1 as $n \rightarrow \infty$, D is bounded between two quantities of the form $cn/2^n$. If n is even and each set is required to contain exactly $n/2$ elements, then, with probability tending to 1, the minimum absolute difference is bounded above by $cn^2/2^n$.

7.4. *Knapsack and subset sum*

The knapsack problem is the zero-one integer programming problem with a single linear constraint. The problem can be described as follows. A set of n items is given. Each item has a specified weight and a specified value. The objective is to select a set of items of maximal total value, such that the sum of the weights of the selected items does not exceed a given bound called the capacity of the knapsack. The knapsack problem is $\mathcal{NP}$-hard, but good worst case approximation algorithms whose execution time is quadratic in the length of the input are known.

The subset sum problem is an ($\mathcal{NP}$-complete) special case of the knapsack problem. The objective is to select a set of items whose total weight is equal to the capacity of the knapsack.

V. Chvátal (1980). Hard knapsack problems. *Oper. Res. 28*, 1402-1411.

A general class of recursive algorithms for the knapsack problem is introduced. These algorithms use the full power of branch-and-bound and dynamic programming as well as rudimentary divisibility arguments. If an n-item instance is generated by drawing the weights independently from the uniform distribution over $\{1,2, \ldots, 10^{n/2}\}$, setting each value equal to the corresponding weight and setting the knapsack capacity equal to half the sum of the weights, then, with probability tending to 1, every recursive algorithm requires at least $2^{n/2}$ steps to solve the instance.

G. d'Atri (1979). *Analyse Probabilistique du Problème du Sac-à-Dos,* Thèse, Université de Paris VI.

The coefficients of the n-variable knapsack problem are assumed to be drawn independently from the uniform distribution over $\{1,2,\ldots,n\}$ and the capacity of the knapsack is drawn from the uniform distribution over $\{1,2,\ldots,cn\}$. A linear time algorithm consisting of a greedy phase followed by an adjustment phase obtains an optimal solution with probability tending to 1.

V. Lifschitz (1980). The efficiency of an algorithm of integer programming: a probabilistic analysis. *Proc. Amer. Math. Soc. 79,* 72-76.

If the coefficients in an n-variable knapsack problem are drawn independently from a common continuous distribution, then a simple enumerative method based on a dominance relation between solutions finds an optimal solution in expected time E_n, where $\ln E_n \simeq 2\sqrt{n}$.

A.M. Frieze, M.R.B. Clarke (1981). *Approximation Algorithms for the m-Dimensional 0-1 Knapsack Problem: Worst-Case and Probabilistic Analyses,* Queen Mary College, University of London.

The integer programming problem $\max\{\Sigma_j c_j x_j : \Sigma_j a_{ij} x_j \leqslant 1,\ i = 1,\ldots,m;\ x_j \in \{0,1\},\ j = 1,\ldots,n\}$ is considered, with m held fixed and n variable. The distribution of the optimal value OPT is considered, and it is shown that a simple rounding algorithm gives a solution of value ROUND, where OPT-ROUND $\leqslant \epsilon$.OPT with probability tending to 1, for all ϵ of the form $n^{-\alpha}$, $\alpha \leqslant 1/(m+1)$.

G.S. Lueker (1982). On the average difference between the solutions to linear and integer knapsack problems. R.L. Disney, T.J. Ott (eds.). *Applied Probability - Computer Science: the Interface, Volume I,* Birkhäuser, Basel, 489-504.

The expected difference between the values of the integer and linear versions of the 0-1 knapsack problem is $O(\log^2 n)$ and $\Omega(1/n)$ when the coefficients are drawn independently from the uniform distribution over [0,1].

A.V. Goldberg, A. Marchetti-Spaccamela (1984). On finding the exact solution of a zero-one knapsack problem. *Proc. 16th Annual ACM Symp. Theory of Computing,* 359-368.

When the (weight, value) pairs are generated by a Poisson process with n as the expected number of items, then, for every $\epsilon > 0$, there is a polynomial time algorithm that solves the knapsack problem to optimality with probability at least $1-\epsilon$. The algorithm depends on the parameters of the Poisson process, and its running time is exponential in $1/\epsilon$.

J.C. Lagarias, A.M. Odlyzko (1983). Solving low-density subset sum problems. *Proc. 24th Annual IEEE Symp. Foundations of Computer Science,* 1-10.

The weights are drawn independently from a uniform distribution over

$\{1, \ldots, 2^{n/d}\}$, and the capacity is the total weight of a randomly chosen subset of items. A polynomial time algorithm is developed that, for 'almost all' problems with $d<1/n$, finds the subset of items whose total weight equals the capacity. The algorithm uses the lattice basis reduction algorithm due to A.K. Lenstra, H.W. Lenstra, Jr., and L. Lovász [*Math. Ann. 261* (1982), 515-534].

A.M. Frieze (1984). *On the Lagarias-Odlyzko Algorithm for the Subset Sum Problem,* Graduate School of Industrial Administration, Carnegie-Mellon University, Pittsburgh, PA.

The above result of Lagarias & Odlyzko is proved in a simpler way and extended to $d<2/n$.

7.5. *Set covering*

The set covering problem is of the form $\min\{e_n x: Ax \geqslant e_m; x_j \in \{0,1\}, j = 1, \ldots, n\}$, where A is an $m \times n$ matrix of zeros and ones and e_h denotes the vector of h ones. A random (m,n,p) problem is one in which the elements of A are independent and each is equal to 1 with probability p.

J.F. Gimpel (1967). A stochastic approach to the solution of large set covering problems. *Proc. 8th Annual IEEE Symp. Switching and Automata Theory,* 76-82.

If $m = \lfloor an \rfloor$ and p is held constant as $n \rightarrow \infty$ then, with probability 1, the ratio between the cost of the optimal solution and the cost of the solution produced by a simple greedy algorithm tends to 1. This is one of the earliest results on the probabilistic analysis of combinatorial algorithms. The result is not true in general if p depends on n (see [Karp 1976] in §1).

C. Vercellis (1984). A probabilistic analysis of the set covering problem. *Ann. Oper. Res. 1.*

Here, p is constant and $m \rightarrow \infty$. If n grows faster than $\log m$ but remains polynomially bounded in M, then the set covering problem is a.s. feasible and the ratio between the optimal solution value and $\log m / \log(1/(1-p))$ tends to 1 a.s. A probabilistic analysis of the asymptotic behavior of two simple heuristics is performed.

8. BRANCH-AND-BOUND AND LOCAL SEARCH

R.M. Karp, J. Pearl (1983). Searching for an optimal path in a tree with random costs. *Artificial Intelligence 21,* 99-116.

Let $F(T)$ be the minimum total weight of a root-leaf path in a tree T. In the case where T is a uniform binary tree of height n and the edge weights are independent Bernoulli random variables with mean p, the distribution of $F(T)$ is studied and polynomial time search strategies are shown to give good near-optimal paths with high probability. The model is proposed as an abstraction of branch-and-bound search.

H. Nakano, Y. Nakanishi (1983). An analysis of local neighborhood search method for combinatorial optimization problems. *Proc. Internat. IEEE Symp. Circuits and Systems,* 1055-1058.

The neighborhood search method of finding locally optimal solutions of combinatorial optimization problems is modeled by a Markov chain. In the case of the λ-opt method for the traveling salesman problem, the predictions of the model are in good agreement with experiment.

D.R. Smith (1984). Random trees and the analysis of branch and bound procedures. *J. Assoc. Comput. Mach. 31,* 163-188.

A model of random branch-and-bound search trees is investigated. The expected time and space complexity of the best bound first and depth first strategies are presented and compared. The results are applied to the traveling salesman problem with random inputs.

C.A. Tovey (1983). On the number of iterations of local improvement algorithms. *Oper. Res. Lett. 2,* 231-238.
C.A. Tovey (to appear). Low order polynomial bounds on the expected performance of local improvement algorithms. *Math. Programming Stud.*
C.A. Tovey (to appear). Hill climbing with multiple local optima. *SIAM J. Algebraic Discrete Methods.*

Tovey's model of local improvement algorithms in combinatorial optimization confirms empirical observations. The model predicts exponential worst case and low order polynomial average performance for problems with a single optimum, such as linear programming and linear complementarity problems. For hard problems with multiple local optima, average speed is linearly bounded but accuracy is poor.

7

Randomized Algorithms

F. Maffioli
Politecnico di Milano

M.G. Speranza, C. Vercellis
Università di Milano

CONTENTS

NOTE. This bibliography has appeared, in a different editorial form, in: F. Archetti, F. Maffioli (eds.) (1984). *Stochastics and Optimization, Ann. Oper. Res. 1,* J.C. Baltzer, Basel.

Our intention has been to provide the reader with a fairly complete list of the existing English literature about randomized algorithms, i.e. algorithms equipped with a *coin tossing* state, with the exception of methods for global continuous optimization and of applications to cryptography.

§1 lists a number of surveys. §2 contains mainly theoretical computer science works. §3 deals with decision problems, i.e., problems requiring a yes-no answer, §4 is concerned with cases in which the use of randomization results into a speed-up of an already efficient (i.e. polynomial) algorithm. §5 is concerned with (discrete) optimization problems, surveying various uses of randomization for them.

The authors wish to acknowledge fruitful discussions with M. Rabin.

1. SURVEYS

The reader interested in a deeper introduction in some aspects of randomized algorithms is invited to look at the following excellent surveys:

G.S. Lueker (1981). Algorithms with random input. W.F. Eddy (ed.). *Proc. 13th Symp. Interface,* Springer, New York.

B.W. Weide (1978). *Statistical Methods in Algorithmic Design and Analysis,* Ph. D. thesis, Report CMU-CS-78-142, Computer Science Department, Carnegie-Mellon University, Pittsburgh, PA.

D.J.A. Welsh (1983). Randomised algorithms. *Discrete Appl. Math. 5,* 133-146.

M.O. Rabin (1977). Complexity of computations, *Comm. ACM 20,* 623-633.

J. Hopcroft (1981). Recent directions in algorithmic research. P. Deussen (ed.). *Proc. 5th Conf. Theoretical Computer Science,* Springer, Berlin, 123-134.

2. Probabilistic Computations: Models and Complexity Classes

The papers of this section are mostly of theoretical nature and are concerned with *classes of languages.* We may in fact consider the following very general recognition problem: given a string x of symbols taken from an alphabet Σ, does x have property π? That is, once we have defined the language L as the set of strings having property π:

$$L = \{x : x \text{ has property } \pi\},$$

we ask, given a certain x: does x belong to L?

Without loss of generality, (a) we may restrict our attention to binary strings (i.e. strings over $\Sigma = \{0,1\}$); (b) x can encode any problem instance; (c) π, the property, may be implicitly defined by asking that the data encoded by x have certain features.

Several models of computation may be utilized to analyze the various languages. The most widely used are *Turing machines.* The *deterministic Turing machine* (DTM) is equipped with a read-only input tape where x is stored, a write-only output tape, a work tape and a finite state automaton equipped with heads for the use of the tapes.

The next step of the machine depends only on the symbol read on the input tape, the symbol on the work tape and the internal state of the automaton, and is entirely deterministic. We say that a DTM M *recognizes* L if and only if M computes the characteristic function $\psi(x)$ of L, that is, M writes 1 if $x \in L$, 0 if $x \notin L$ on its output tape and halts after a finite number m of steps for any input x.

If there exists a polynomial p such that $m \leqslant p(|x|)$ for all x, we say that M recognizes L in polynomial time, and that $L \in \mathcal{P}$.

We may think of a much more powerful machine, the *nondeterministic Turing machine,* NTM, as a machine equipped with a *guessing* 'black box' which, given x, produces another string y (a *certificate*) from which a DTM is able to answer the question $x \in L$ in polynomial time, whereas it is not known how to do the same without y. Notice that $|y|$ must be polynomially bounded as well: this class of languages is known as $\mathcal{NP}$.

Consider finally a machine equipped with a *coin tossing* black box which at each step flips an unbiased coin and uses this random result together with the other elements at its disposal to decide on the next step: in this case we have a *probabilistic Turing machine* (PTM), which was considered for the first time in

K. de Leeuw, E.F. Moore, C.E. Shannon, N. Shapirio (1955). Computability by probabilistic machines. C.E. Shannon, J. McCarthy (eds.). *Automata Studies,* Princeton University Press, Princeton, NJ, 183-212.

In general a PTM M computes a random output string y with probability $Pr\{M(x) = y\}$. Note also that, once the string z of the results of the coin tossing black box is known, we obtain an entirely deterministic machine M', with $M'(x,z) = y$.

Assume now that there are input strings x, for which among all outputs one, say y, is obtained at least half of the times, i.e., $Pr\{M(x) = y\} > ½$. Then we say that M *computes* a partial function $\phi(x)$, which is equal to y whenever $Pr\{M(x) = y\} > ½$ and is undefined otherwise. The *error probability* of M is then the function $e(x)$, which equals $Pr\{M(x) \neq \phi(x)\}$ whenever $\phi(x)$ is defined and is undefined otherwise. Obviously $e(x) < ½$, whenever it is defined, but for a PTM to be useful it must exhibit a *bounded error probability,* i.e., there must exist an $\epsilon < ½$ such that $e(x) \leq \epsilon$ for every x in the domain of ϕ. The crucial difference between the two types of PTM is better appreciated if one considers that in the second case, by repeating the computation for the same input x a sufficiently large number of times, the probability of error may be reduced as much as we wish, whereas this is not in general possible with the first machine. In fact, by taking the majority decision on whether to accept or not x as a member of L, the probability of error after $2N+1$ trials is reduced below $\Sigma_{k=0}^{N} \binom{2N+1}{k} \epsilon^{2N-k+1}(1-\epsilon)^k$. When $\epsilon < ½$, this sum approaches zero at an exponential rate when $N \rightarrow \infty$ whereas this is not the case if we require only that $e(x) < ½$; consider for instance $e(x) = ½ - (½)^{|x|}$. These definitions were introduced in

J.T. Gill (1977). Computational complexity of probabilistic Turing machines. *SIAM J. Comput. 6,* 675-695.

The PTM considered here corresponds to what Rabin would call a *probabilistic autmaton,* with an *isolated cutpoint* at ½; see the pioneering paper

M.O. Rabin (1963). Probabilistic automata. *Inform. and Control 6,* 230-245.

Other quite early references are;

E.S. Santos (1969). Probabilistic Turing machines and computability. *Proc. Amer. Math. Soc. 22,* 704-710.

E.S. Santos (1971). Computability by probabilistic Turing machines. *Trans. Amer. Math. Soc. 4,* 165-184.

The reader can also look at the following more recent papers:

G.J. Chaitin (1977). Algorithmic information theory. *IBM J. Res. Develop. 21,* 350-359.

P.J. Davis (1977). Proof, completeness, transcendentals and sampling. *J. Assoc. Comput. Mach. 24,* 298-310.

R. Freivalds (1977). Probabilistic machines can use less running time. B. Gilchrist (ed.). *Information Processing 77,* North-Holland, Amsterdam, 839-842.

A.C. Yao (1977). Probabilistic computations: towards a unified measure of complexity. *Proc. 18th Annual IEEE Symp. Foundations of Computer Science,* 222-227.

Similarly as for DTM or NTM, we say that a PTM *recognizes* a language L if it computes the characteristic function of that language. The *running time* $t(x)$ of a PTM M is defined as the least positive integer n such that $Pr\{M(x) = \phi(x) \text{ in } n \text{ steps}\} > 1/2$ if ϕ is defined and ∞ when ϕ is undefined. A PTM (or NTM) M is *polynomial bounded* if for all x it halts after a number of steps bounded from above by a polynomial in $|x|$, the length of the input. We shall call:
- $\mathcal{PP}$ the class of languages recognized by polynomial bounded PTMs;
- $\mathcal{BPP}$ the class of languages recognized by polynomial bounded PTMs with bounded error probability;
- $\mathcal{RP}$ the class of languages recognized by polynomial bounded PTMs which have zero probability of error for inputs in the complements of the languages.

Recall that $\mathcal{P}$SPACE denotes the class of languages recognized by DTMs in polynomial space, i.e., utilizing a polynomial bounded number of cells of the work tape.

In [Gill 1977] it is shown that $\mathcal{P} \subseteq \mathcal{RP} \subseteq \mathcal{BPP} \subseteq \mathcal{PP} \subseteq \mathcal{P}$SPACE and that $\mathcal{RP} \subseteq \mathcal{NP} \subseteq \mathcal{PP}$, none of these inclusions being known to be proper. In

Ker-I Ko (1982). Some observations on the probabilistic algorithms and NP-hard problems. *Inform. Process. Lett. 14,* 39-43

it is proved that $\mathcal{NP} \subseteq \mathcal{BPP}$ is equivalent to $\mathcal{RP} = \mathcal{NP}$. Unlike $\mathcal{RP}$, both $\mathcal{PP}$ and $\mathcal{BPP}$ are closed under complementation, and in fact one can easily show that $\mathcal{RP} \cup$ co-$\mathcal{RP} \subseteq \mathcal{BPP}$.

In order to survey some other recent results we need to introduce the concept of TMs with oracles. An *oracle* for a specified language L is a 'black box' that, given a string x as input, will tell in one step whether or not $x \in L$. A TM provided with an oracle for the language L is obviously more powerful than without it. We say that the class of languages recognized by such a machine is *relativized to* L and indicate this by writing L as a superscript. For instance $\mathcal{P}^{SAT}$ is the class of languages recognized in polynomial time by a DTM equipped with an oracle which recognizes in unit time if a string x encodes a satisfiable boolean function in conjunctive normal form (i.e. an oracle solving in unit time the satisfiability problem). If $\mathcal{C}$ is a collection of languages, we define $\mathcal{P}^{\mathcal{C}} = \cup_{L \in \mathcal{C}} \mathcal{P}^{L}$ and $\mathcal{NP}^{\mathcal{C}} = \cup_{L \in \mathcal{C}} \mathcal{NP}^{L}$. These definitions are utilized in

L.J. Stockmeyer (1977). The polynomial-time hierarchy. *Theoret. Comput. Sci. 3*, 1-22

to introduce recursively the *polynomial hierarchy* as follows:

$$\Sigma_0^p = \Pi_0^p = \Delta_0^p = \mathcal{P},$$

and for $k \geqslant 0$:

$$\Sigma_{k+1}^p = \mathcal{NP}^{\Sigma_k^p},$$
$$\Pi_{k+1}^p = \{A : \overline{A} \in \Sigma_{k+1}^p\} = \text{co-}\Sigma_{k+1}^p,$$
$$\Delta_{k+1}^p = \mathcal{P}^{\Sigma_k^p}.$$

We also let $\mathcal{PH} = \cup_{k=0}^{\infty} \Sigma_k^p$. It is known that $\mathcal{PH} \subseteq \mathcal{P}$SPACE. Note that this whole hierarchy collapses (i.e. $\mathcal{PH} = \mathcal{P}$) if the famous $\mathcal{P} =? \mathcal{NP}$ question is given an affirmative answer.

As far as the hope of answering the $\mathcal{P} =? \mathcal{NP}$ question, a (negative) result can be found in

T. Baker, J. Gill, R. Solovay (1975). Relativization of the P=?NP question. *SIAM J. Comput 4*, 431-442.

Here it is shown that there are oracles B and C such that both $\mathcal{P}^B = \mathcal{NP}^B$ and $\mathcal{P}^C \neq \mathcal{NP}^C$. For more about this see

C.H. Bennett, J. Gill (1981). Relative to a random oracle A, $\mathrm{P}^{\mathrm{A}} \neq \mathrm{NP}^{\mathrm{A}} \neq$ co-NP^{A} with probability 1. *SIAM J. Comput. 10*, 96-113.

This is a recent effort to cast new light on the question. Oracles such as B above are shown to be rare, i.e., *almost always* given an oracle A chosen at random among all possible ones $\mathcal{P}^A \neq \mathcal{NP}^A$. This would solve in the negative (as conjectured) the $\mathcal{P} =? \mathcal{NP}$ question if one could prove the *random oracle hypothesis*: 'an unrelativized statement $\mathbb{S}$ is true if and only if a corresponding *acceptably* relativized statement $\mathbb{S}^A$ is almost always true, where A is a random oracle'. However, this hypothesis, at least as formulated above, has been proved false in

S.A. Kurtz (1982). On the random oracle hypothesis. *Proc. 14th Annual ACM Symp. Theory of Computing*, 224-230,

where a pair of classes is exhibited for which an unrelativized statement is true whereas the corresponding acceptably relativized statement with respect to a random oracle A is false with probability 1.

The class $\mathcal{RP}$ is another candidate for characterizing more sharply the frontier between tractable and intractable problems. The question $\mathcal{P} =? \mathcal{NP}$ can therefore be replaced by the $\mathcal{RP} =? \mathcal{NP}$ question. Against an affirmative answer to this it has been shown that $\mathcal{NP} \neq \mathcal{RP}$ unless again the whole polynomial hierarchy collapses, since in that case $\mathcal{PH} = \Delta_2^p$. See

R.M. Karp, R. Lipton (1980). Some connections between non-uniform and uniform complexity classes. *Proc. 12th Annual ACM Symp. Theory of Computing*, 302-309.

A result similar to that of [Baker, Gill & Solovay 1975] (see above) in pointing out the difficulty of the $\mathcal{RP} =? \mathcal{NP}$ question is found in

C. Rackoff (1982). Relativized questions involving probabilistic algorithms. *J. Assoc. Comput. Mach. 29*, 261-268:

there are oracles A and B such that both $\mathcal{P}^A = \mathcal{RP}^A \neq \mathcal{NP}^A$ and $\mathcal{P}^B \neq \mathcal{RP}^B = \mathcal{NP}^B$.

Ker-I Ko enhances in [Ker-I Ko 1982] (see above) the key role of $\mathcal{RP}$ by proposing the following hierarchy:

$$\Sigma_0^r = \Pi_0^r = \Delta_0^r = \mathcal{P},$$

and for $k \geqslant 0$

$$\Sigma_{k+1}^r = \mathcal{RP}^{\Sigma_k^r}, \quad \Pi_{k+1}^r = \text{co-}\Sigma_{k+1}^r, \quad \Delta_{k+1}^r = \mathcal{P}^{\Sigma_k^r}.$$

Then if we let $\mathcal{RH} = \cup_{k=0}^{\infty} \Sigma_k^r$, we have from $\mathcal{RP} \subseteq \mathcal{NP}$ that $\mathcal{RH} \subseteq \mathcal{PH}$. It is not difficult to prove that $\mathcal{NP} \not\subseteq \mathcal{RH}$ unless $\mathcal{RP} = \mathcal{NP}$ and therefore the exact relation between $\mathcal{RH}$ and $\mathcal{PH}$ also depends on the question whether $\mathcal{RP} \subseteq \mathcal{NP}$ is a proper inclusion or not.

It is believed that added evidence against the possibility that $\mathcal{RP} = \mathcal{NP}$ is given in

L. Adleman, K. Manders (1977). Reducibility, randomness and intractability. *Proc. 9th Annual ACM Symp. Theory of Computing*, 151-163,

L. Adleman (1980). Two theorems on random polynomial time. *Proc. 19th Annual IEEE Symp. Foundations of Computer Science*, 75-83.

Here it is shown that, if a language L belongs to $\mathcal{RP}$, it has a *polynomial size circuit*, which means that it is possible to build a logic circuit with a number of Boolean gates bounded by a polynomial in $|x|$ (the size of the strings we are testing) to compute a function $L_{|x|}(x)$ such that

$$L_{|x|}(x) = \begin{cases} 1 & \text{if } x \in L, \\ 0 & \text{otherwise.} \end{cases}$$

We close this section by recalling that in

J. Simon (1977). On the difference between the one and the many. A. Salomaa, M. Steinby (eds.). *Automata, Languages and Programming*, Lecture Notes in Computer Science 52, Springer, Berlin, 480-491,

the class $\mathcal{PP}$ is shown practically to coincide with the class of *counting problems* whose complexity has been investigated in

B.A. Trakhtenbrot (1975). On problems solvable by successive trials. J. Bečvář (ed.). *Mathematical Foundations of Computer Science 1975,* Lecture Notes in Computer Science 32, Springer, Berlin, 125-137.

L.G. Valiant (1979). The complexity of enumeration and reliability problems. *SIAM J. Comput. 8,* 410-440.

A typical complete (i.e. hardest) problem for this class is the following: 'does the graph G have at least k distinct Hamiltonian cycles?' No hardest problem is known for $\mathcal{BPP}$.

Other relevant contributions to this area of theoretical computer science not explicitly mentioned before are:

R. Aleliunas, R. Karp, R. Lipton, L. Lovász, C. Rackoff (1979). Random walks, universal sequences and the complexity of maze problems. *Proc. 20th Annual IEEE Symp. Foundations of Computer Science,* 218-223.

J.M. Barzdin (1970). On the frequency solution of algorithmically unsolvable mass problems. *Dokl. Akad. Nauk. SSSR 191,* 967-970.

R. Freivalds (1978). Recognition of languages with high probability of different classes of automata. *Soviet Math. Dokl. 19,* 295-298.

R. Freivalds (1981). Projections of languages recognizable by probabilistic and alternating finite multitape automata. *Inform. Process. Lett. 13,* 195-198.

R. Freivalds (1981). Probabilistic two-way machines. J. Gruska, M. Chytil (eds.). *Mathematical Foundations of Computer Science 1981,* Lecture Notes in Computer Science 118, Springer, Berlin, 33-45.

H. Jung (1981). Relationships between probabilistic and deterministic tape complexity. *Ibidem,* 339-346.

D. Kozen (1979). Semantics of probabilistic programs. *Proc. 20th Annual IEEE Symp. Foundations of Computer Science,* 104-114.

R. Ladner (1975). On the structure of polynomial time reducibility. *J. Assoc. Comput. Mach. 22,* 155-171.

D.J. Lehmann, M.O. Rabin (1981). On the advantage of free choice: a symmetric and fully distributed solution to the dining philosophers problems. *Proc. 8th ACM Symp. Principles of Programming Languages,* 133-138.

M.O. Rabin (1979). *Probabilistic Algorithms,* Report RC 6164, IBM Research Division, San Jose-Yorktown-Zürich.

J.H. Reif (1980). Logics for probabilistic programming. *Proc. 12th Annual ACM Symp. Theory of Computing,* 8-13.

W.L. Ruzzo, J. Simon, M. Tompa (1982). Space-bounded hierarchies and probabilistic computations. *Proc. 14th Annual ACM Symp. Theory of Computing,* 215-223.

J. Simon (1981). Space-bounded probabilistic Turing machine complexity classes are closed under complement. *Proc. 13th Annual ACM Symp. Theory of Computing,* 158-167.

3. Decision Problems

3.1. *Primality*

Several probabilistic algorithms have been succesfully applied to the problem of deciding whether a given number n is prime or not. In

M.O. Rabin (1980). Probabilistic algorithms for testing primality. *J. Number Theory 12,* 128-138,

R. Solovay, V. Strassen (1977;1978). A fast Monte-Carlo test for primality. *SIAM J. Comput 6,* 84-85; Erratum. *SIAM J. Comput. 7,* 118,

it is shown that primality belongs to the class $\mathcal{RP}$; this is a quite important result as this is the only famous problem which is not known either to belong to $\mathcal{P}$ or to be $\mathcal{NP}$-complete, but which has been proved to belong to $\mathcal{RP}$. In both the algorithms presented in these papers, the output 'n is prime' has a probability of being incorrect which is less than ¼ for the Rabin algorithm and less than ½ for the Solovay-Strassen algorithm.

A survey on primality is presented in

H.C. Williams (1978). Primality testing on a computer. *Ars Combin. 5,* 127-185,

where it is proved that the Solovay-Strassen algorithm is a weaker form of the Rabin test. Moreover, this paper presents a probabilistic algorithm for primality due to Malm, which gives the output 'n is prime' with probability of error less than $2^{-(m-1)}$, where m is the number of distinct prime factors of n, and which is not very efficient. In

A.O.L. Atkin, R.G. Larson (1982). On a primality test of Solovay and Strassen. *SIAM J. Comput. 11,* 789-791,

it is proved that the strong pseudo-prime test is better than the Solovay-Strassen algorithm, in the sense that it never takes more time nor is less effective. The probability of error in the strong pseudo-prime test is shown to be never greater than ¼. A different from of the Rabin test is presented in

T. Harlestam (1980). A note on Rabin's probabilistic primality test. *BIT 20,* 518-521.

Results of theoretical interest in

G.J. Chaitin, J.T. Schwartz (1978). A note on Monte-Carlo primality tests and algorithmic information theory. *Comm. Pure Appl. Math. 31,* 521-527

show how to obtain error-free tests for primality from the Rabin and Solovay-Strassen algorithms.

Both the possible outputs of the probabilistic algorithm for primality presented in

L. Adleman (1980). On distinguishing prime numbers from composite numbers. *Proc. 21st Annual IEEE Symp. Foundations of Computer Science*, 387-406

are always correct, but it is possible (with small probability) for the algorithm not to terminate.

A different algorithm, which is very simple and efficient, is presented in

D.J. Lehmann (1982). On primality tests. *SIAM J. Comput. 11*, 374-375.

Both conclusions 'n is prime' and 'n is composite' may be incorrect, with probability less than ½.

3.2. *Other decision problems*

Most of the probabilistic algorithms presented for decision problems are similar to the Rabin and Solovay-Strassen algorithms for primality, with one answer always correct and the other possibly incorrect with probability less than ½. The algorithm presented in

L. Babai (1979). *Monte-Carlo Algorithms in Graph Isomorphism Testing*, Rapport de recherches DMS 79-10, Département de Mathématique et de Statistique, Université de Montréal

for graph isomorphism testing is slightly but essentially different, in the sense that there are three possible outputs: 'yes', 'no', and '?' (failure symbol), the first two being always correct and the probability of the third one being less than ½. For this kind of algorithm Babai proposed the term 'Las Vegas', reserving the term 'Monte Carlo' for the Rabin type algorithms.

Monte Carlo algorithms for the verification of polynomial identities and properties of systems of polynomials have been presented in

J.T. Schwartz (1980). Fast probabilistic algorithms for verification of polynomial identities. *J. Assoc. Comput. Mach. 27*, 701-717.

Algorithms for testing of multiplication of polynomials, as well as algorithms for multiplication of integers and matrices can be found in

R. Freivalds (1979). Fast probabilistic algorithms. J. Bečvář (ed.). *Mathematical Foundations of Computer Science 1979*, Lecture Notes in Computer Science 74, Springer, Berlin, 57-69.

For the problem of deciding whether an n-variable polynomial F is absolutely prime a Monte Carlo algorithm, polynomial in n and exponential in the degree of F, is presented in

J. Heintz, M. Sieveking (1981). Absolute primality of polynomials is decidable in random polynomial time in the number of variables. S. Even, O. Kariv (eds.). *Automata, Languages and Programming,* Lecture Notes in Computer Science 115, Springer, Berlin, 16-28.

A further example of the importance of efficient probabilistic algorithms, also for problems in $\mathcal{P}$, is given in

L. Lovász (to appear). *On Determinants, Matchings and Random Algorithms,* Unpublished manuscript.

Among other results, Lovász formulates Tutte's necessary and sufficient condition for the existence of a perfect matching and reduces the problem to verifying whether the determinant of a given matrix is identically 0, which can be done efficiently with Schwartz's probabilistic algorithms mentioned above. Lovász presents analogous results for a generalization of this problem called the matroid matching problem.

The rank decision problem, for which no polynomial algorithm is known, is shown to be solvable probabilistically in polynomial time, as well as the problem of deciding equivalence of straight line programs, in

Y. Yemini (1979). *On Some Randomly Decidable Geometrical Problems,* Unpublished manuscript,

O.H. Ibarra, S. Moran (1981). Deterministic and probabilistic algorithms for maximum bipartite matching via fast matrix multiplications. *Inform. Process. Lett. 13,* 12-15,

O.H. Ibarra, S. Moran (1981). Probabilistic algorithms for deciding equivalence of straight line programs. *J. Assoc. Comput. Mach. 30,* 217-228.

The problem of deciding whether two free Boolean graphs (representations of a special class of Boolean functions) are equivalent, is shown to belong to $\mathcal{RP}$ in

M. Blum, A.K. Chandra, M.N. Wegman (1980). Equivalence of free boolean graphs can be decided probabilistically in polynomial time. *Inform. Process. Lett. 10,* 80-82.

A deterministic polynomial algorithm for this problem is not known.

For more in this area see also:

W. Janko (1980). Probabilistic algorithms and the efficient use of statistical models of decision in their design. *Math. Oper. Res. 36,* 165-182;

K. Mehlhorn, E.M. Schmidt (1982). Las Vegas is better than determinism in VLSI and distributed computing. *Proc. 14th Annual ACM Symp. Theory of Computing,* 330-337;

G.L. Miller (1975). Riemann's hypothesis and tests for primality. *Proc. 7th Annual ACM Symp. Theory of Computing,* 234-239;

J.M. Pollard (1975). A Monte-Carlo method for factorization. *BIT 15,* 331-334;

M.O. Rabin (1981). *Fingerprinting by Random Polynomials,* Research report TR-15-81, Aiken Computation Laboratory, Harvard University, Cambridge, MA;

Y. Yacobi, S. Even (1980). *A 'Hard-Core' Theorem for Randomized Algorithms,* Technical report 188, Computer Science Department, Technion, Haifa;

A.C. Yao (1977). *A Lower Bound on Palindrome Recognition by Probabilistic Turing Machines,* Report STAN-CS-77-647, Computer Science Department, Stanford University, Stanford, CA.

4. Speed-Up Via Probabilistic Algorithms

Various exact probabilistic algorithms have been presented for solving different kinds of problems. They achieve a speed-up if their expected running time (the average over the number of inner random steps) results in a reduction of the best known deterministic running time.

A typical example of probabilistic speed-up is the probabilistic version of the quicksort algorithm for sorting a list of n elements:

C.A.R. Hoare (1962). Quicksort. *Comput. J. 5,* 10-15,

R. Sedgewick (1977). The analysis of quicksort programs. *Acta Inform. 7,* 327-355.

The deterministic quicksort has a $O(n^2)$ worst-case deterministic running time; when the partitioning element is randomly chosen it results in a probabilistic quicksort algorithm whose expected running time is $O(n \log n)$, which is also the best deterministic running time for the sorting problem. Another probabilistic variant of quicksort is presented in

W.D. Frazer, A.C. McKellar (1970). Samplesort: a sampling approach to minimal storage tree sorting. *J. Assoc. Comput. Mach. 17,* 496-507.

Deterministic variants of quicksort, which easily result in probabilistic algorithms, are presented in

R.C. Singleton (1969). Algorithm A 347: an efficient algorithm for sorting with minimal storage. *Comm. ACM 12,* 185-187

and in [Sedgewick 1977] above.

Sorting in networks have also been successfully approached by a randomized algorithm in

J.H. Reif, L.G. Valiant (1983). A logarithmic sort for linear size networks. *Proc. 15th Annual ACM Symp. Theory of Computing,* 10-16.

An exact probabilistic algorithm for checking equivalence of circular lists is presented and shown to be analogous to probabilistic quicksort in

A. Itai (1979). A randomized algorithm for checking equivalence of circular lists. *Inform. Process. Lett. 9,* 118-121.

In fact, using a simple initial random rotation of the lists reduces the deterministic $3n-3$ running time to a $2n-1$ expected running time.

A probabilistic speed-up algorithm for finding the nearest pair among n points in a k-dimensional space is presented in

M.O. Rabin (1976). Probabilistic algorithms. J. Traub (ed.). *Algorithms and Complexity: New Directions and Recent Results,* Academic Press, New York, 21-39.

It reduces the $O(n \log n)$ best known deterministic running time to an $O(n)$ expected time. A comment on this algorithm appears in

S. Fortune, J. Hopcroft (1979). A note on Rabin's nearest neighbour algorithm. *Inform. Process. Lett. 8,* 20-23.

For the problem of computing a minimum spanning tree for n points in a p-dimensional space a probabilistic algorithm is presented in

F.J. Rohlf (1978). A probabilistic minimum spanning tree algorithm. *Inform. Process. Lett. 7,* 44-48.

Its expected running time is conjectured to be $O(n \log\log n)$.

A probabilistic algorithm for computing the vertex connectivity of a graph in expected time $O((-\log\epsilon)|V|^{3/2}|E|)$ with error probability not greater than ϵ is presented in

M. Becker, W. Degenhardt, J. Doenhardt, S. Hertel, G. Kaninke, W. Weber, K. Mehlhorn, S. Naher, H. Rohnert, T. Winter (1982). A probabilistic algorithm for vertex connectivity of graphs. *Inform. Process. Lett. 15,* 135-136.

For the problem of finding the maximum matching in a bipartite graph an efficient probabilistic algorithm turns out to have error probability less than ½ in

O.H. Ibarra, S. Moran (1981). Deterministic and probabilistic algorithms for maximum bipartite matching via fast matrix multiplication. *Inform. Process. Lett. 13,* 12-15.

A probabilistic algorithm for covering a graph by circuits takes $O(|V|^2)$ time on the average and may not terminate with probability zero:

A. Itai, M. Rodeh (1978). Covering a graph by circuits. G. Ausiello, C. Böhm (eds.). *Automata, Languages and Programming,* Lecture Notes in Computer Science 62, Springer, Berlin, 289-299.

Probabilistic algorithms for problems pertaining to finite fields, like generating an irreducible polynomial of degree n, finding a root of a polynomial and factoring a given polynomial, are presented in

E.R. Berlekamp (1970). Factoring polynomials over large finite fields, *Math.*

Comp. 24, 713-735,

M.O. Rabin (1980). Probabilistic algorithms in finite fields. *SIAM J. Comput. 9,* 273-280.

An improvement of Rabin's algorithm for generating irreducible polynomials, together with some computational results, is presented in

J. Calmet, R. Loos (1980). An improvement of Rabin's probabilistic algorithm for generating irreducible polynomials over $GF(p)$. *Inform. Process. Lett. 11,* 94-95.

A probabilistic variation of the modular GCD algorithm which is efficient for sparse polynomials is presented in

R.E. Zippel (1979). Probabilistic algorithms for sparse polynomials. E.W. Ng (ed.). *Symbolic and Algebraic Manipulation,* Lecture Notes in Computer Science 72, Springer, Berlin, 216-226.

Other relevant papers in this area are:

J.L. Carter, M.N. Wegman (1977). Universal classes of hash functions. *Proc. 9th Annual ACM Symp. Theory of Computing,* 106-112;

M.L. Fredman (1975). Two applications of a probabilistic search technique: sorting $x+y$ and building balanced search trees. *Proc. 7th Annual ACM Symp. Theory of Computing,* 240-244;

T. Leighton, M. Lepley (1982). Probabilistic searching in sorted linked lists. *Proc. 20th Allerton Conf. Communication, Control, and Computing,* 500-506;

N. Pippenger (1982). Probabilistic simulation. *Proc. 14th Annual ACM Symp. Theory of Computing,* 17-26;

J.H. Reif (1982). On the power of probabilistic choice in synchronous parallel computations. M. Nielsen, E.M. Schmidt (eds.). *Automata, Languages and Programming,* Lecture Notes in Computer Science 140, Springer, Berlin, 442-450;

R. Reischuk (1981). A fast probabilistic parallel sorting algorithm. *Proc. 22nd Annual IEEE Symp. Foundations of Computer Science,* 212-219.

5. Optimization Problems

The use of randomization principles in the design of algorithms has been proposed, since the mid-sixties, as a practical technique to cope with the overwhelming computational intractability of $\mathcal{NP}$-hard combinatorial optimization problems.

Each execution of a *randomized approximation algorithm* (RAA) for a combinatorial optimization problem P results, for any given instance of P, in a random feasible solution, which depends on the outcomes of the random steps incorporated in the RAA. Thus, RAAs are usually embedded, as basic sampling techniques, into Monte Carlo schemes: repeated independent 'trials' of the RAA on the given instance are performed, and the best observed solution value is retained as an approximation to the optimum value.

Three main questions arise in connection with this approach: (i) how to design effective RAAs for generating 'good' solution values; (ii) how to devise a stopping rule for controlling the repeated trials; and (iii) how to evaluate the quality of the solution (e.g. in terms of the ratio of the approximate value to the optimum one).

The earlier, and perhaps the most popular, proposed approach relies on *local search* techniques. Given a minimization (maximization) problem with objective function f and feasible region F, RAAs based on local search schemes require that, with each point $x \in F$, there is associated a predefined *neighborhood* $N(x) \subset F$. Given a current point $x_i \in F$, the set $N(x_i)$ is searched for a point x_{i+1} with $f(x_{i+1}) < f(x_i)$ ($f(x_{i+1}) > f(x_i)$). If such a point exists, it becomes the new current solution, and the process is iterated. Otherwise, x_i is retained as a *local optimum* with respect to $N(x_i)$. Then, a set of feasible solutions is randomly generated, and each of them is 'locally' improved until a solution which is minimum within its neighborhood is reached. To apply the resulting RAA to a particular problem, one has to specify the structure of the neighborhoods and the randomized procedure for obtaining the starting feasible solutions. In

S. Lin (1965). Computer solutions of the traveling salesman problem. *Bell System Tech. J. 44,* 2245-2269,

S. Lin, B.W. Kernighan (1973). An effective heuristic for the traveling salesman problem. *Oper. Res. 21,* 498-516,

R. Roth (1969). Computer solutions to minimum-cover problems. *Oper. Res. 17,* 455-465,

R. Roth (1970). An approach to solving linear discrete optimization problems. *J. Assoc. Comput. Mach. 17,* 303-313,

K. Steiglitz, P. Weiner, D.J. Kleitman (1969). The design of minimal cost survivable networks. *IEEE Trans. Circuits and Systems CT-16,* 455-460,

B.W. Kernighan, S. Lin (1970). An efficient heuristic procedure for partitioning graphs. *Bell System Tech. J. 49,* 291-307,

different local-search based RAAs have been proposed, respectively, for the traveling salesman problem, the set covering problem, the general 0-1 programming problem, the minimum-cost survivable network problem, and the graph partitioning problem. Other applications of this approach are referenced in

C.H. Papadimitriou, K. Steiglitz (1982). *Combinatorial Optimization: Algorithms and Complexity,* Prentice Hall, Englewood Cliffs, NJ.

A different local search scheme, motivated by analogy with statistical mechanics techniques, has recently been proposed in

S. Kirkpatrick, C.D. Gelat, Jr., M.P. Vecchi (1983). Optimization by simulated annealing. *Science 220,* 671-680,

V. Cěrny (1982). *A Thermodynamical Approach to the Travelling Salesman Problem: an Efficient Simulation Algorithm,* Report, Institute of Physics and Biophysics, Comenius University, Bratislava,

B.M. Schwartzschild (1982). Statistical mechanics algorithm for Monte Carlo optimization. *Physics Today 35,* 17-19.

The basic idea is that of generating a 'random exchange' in the current feasible solution, accepting the modified solution as new current not only if the value of the objective function is better, like in the usual local search, but also if it is worse, with a probability given by a Boltzmann-like law, in which the temperature becomes a tunable parameter for the control of the repeated trials. An application of the above technique to the quadratic assignment problem is presented in

R.E. Burkard, F. Rendl (to appear). A thermodynamically motivated simulation procedure for combinatorial optimization problems. *European J. Oper. Res.*

It has been suggested in

P.M. Camerini, A. Colorni, F. Maffioli (to appear). Some experience in applying a stochastic method to location problems. *Math. Programming Stud.*

that clustering-and-sampling techniques, successfully applied in the recent past to global optimization problems, could achieve significant computational savings when applied to combinatorial optimization problems.

The questions of controlling the repeated trials and evaluating the solution quality for local search based RAAs have been addressed by a number of authors:

B.L. Golden (1977). A statistical approach to the TSP. *Networks 7,* 209-225;

B.L. Golden, F.A. Alt (1979). Interval estimation of a global optimum for large combinatorial problems. *Naval Res. Logist. Quart. 26,* 69-77;

C. Lardinois (1979). *Méthodes Statistiques en Optimization Combinatoire,* Rapport 116, Centre de recherche sur les transports, Université de Montréal;

K.L. McRoberts (1971). A search model for evaluating combinatorially explosive problems. *Oper. Res. 19,* 1331-1349.

These papers rely on the use of results from the asymptotic theory of extreme order statistics, and lead to a number of different confidence intervals for the unknown optimum value, postulating a Weibull distribution of the sampled extrema. A similar statistical approach is applied in

D.G. Dannenbring (1977). Procedures for estimating optimal solution values for large combinatorial problems. *Management Sci. 23,* 1273-1283

to a class of more general RAAs (not necessarily local search based). Different statistical techniques are presented in

S. Reiter, G. Sherman (1965). Discrete optimizing. *SIAM J. 13,* 864-889,

S.L. Savage (1976). Some theoretical implications of local optimization. *Math. Programming 10,* 354-366

for the control of local search based RAAs. An alternative approach to the control of a general class of RAAs has been proposed in

C. Vercellis (1981). *A Probabilistic Stopping Rule for Randomized Algorithms,* Research report 81.2, IAMI/CNR, Milan

through a stopping rule in a Bayesian decision theoretic framework. This latter approach has been applied, in particular, to a class of RAAs which follow a randomized greedy strategy: the element to enter the solution at a given step is not the one of maximum (minimum) 'weight', but instead it is randomly chosen according to a probability mass function which attributes higher probability to elements of larger weight. The idea of randomizing a heuristic (and deterministic) algorithm is quite general, and can be applied in order to tackle the pathological effect of worst-case instances.

Finally, we remark that it has been possible to analyze probabilistically the performance of some simple RAAs assuming a distribution over the instances of the given problem. For instance, the simple 'random extension' algorithm for the maximum clique problem is analyzed in

G.R. Grimmett, C.J.H. McDiarmid (1975). On colouring random graphs. *Math. Proc. Cambridge Philosoph. Soc. 77,* 313-324,

R.M. Karp (1976). The probabilistic analysis of some combinatorial search algorithms. J.F. Traub (ed.). *Algorithms and Complexity: New Directions and Recent Results,* Academic Press, New York, 1-19,

under different probabilistic assumptions. A more sophisticated 'extension rotation' randomized algorithm is considered in

L. Posa (1976). Hamiltonian circuits in random graphs. *Discrete Math. 14,* 359-364,

J.H. Reif, P.G. Spirakis (1981). *Probabilistic Analysis of Random Extension-Rotation Algorithms,* Technical report TR-28-81, Aiken Computation Laboratory, Harvard University, Cambridge, MA,

respectively for determining a Hamiltonian path and for the general problem of determining a set of maximum cardinality in an independence system. In

D. Angluin, L.G. Valiant (1979). Fast probabilistic algorithms for Hamiltonian circuits and matchings. *J. Comput. System Sci. 18,* 155-193,

RAAs with both extension and rotation operations are analyzed, for the perfect matching problem and the Hamiltonian circuit problem in a directed graph. Two simple randomized algorithms for the set covering problem are probabilistically analyzed in

C. Vercellis (1984). A probabilistic analysis of the set covering problem. *Ann. Oper. Res. 1*.

8

Parallel Algorithms

G.A.P. Kindervater, J.K. Lenstra
Centre for Mathematics and Computer Science
Amsterdam

CONTENTS

Parallel computing is receiving a rapidly increasing amount of attention. In theory, a collection of processors that operate in parallel can achieve substantial speedups. In practice, technological developments are leading to the actual construction of such devices at low cost. Given the inherent limitations of traditional sequential computers, these prospects turn out to be very stimulating for researchers interested in the design of computers and algorithms.

In this bibliography, we have tried to collect the literature on parallel computing that is relevant for the mathematics of operations research, in particular for the theory of combinatorial optimization. Its organization is as follows.

§1 is concerned with *machine models* designed for parallel computation.

Rather than including the complete literature on this topic, which could fill a sizeable bibliography by itself, we have only surveyed papers that are of general interest or that define models referred to in later sections. Many of the references in §4 and §5 mention machine models for which specific results have been obtained, and the reader who is interested in the characteristics of, say, an *SIMD machine with shared memory, simultaneous reads and no simultaneous writes* should consult §1.

§2 deals with the *complexity theory* of parallel computation. Beyond the basic distinction between *solvability in polynomial time* and *completeness for* $\mathcal{NP}$ in sequential computation, many concepts have been defined and analyzed that are relevant for parallel computing. Again, we have not aimed at a complete survey of this area, but important notions like *solvability in polylog parallel time* and *log space completeness for* $\mathcal{P}$ are introduced here.

§3 gives results for *numerical problems.* Problems like evaluating arithmetic expressions and recurrence relations, solving systems of linear equations and computing eigenvalues have been subjected to parallelization earlier and more extensively than combinatorial problems (*l'histoire se répète*: floating point arithmetic was well understood before anyone had heard of the traveling salesman). §§3.1−2 list references on those subjects, without much comment. §3.3 contains three papers on *nonlinear optimization,* an area in which parallel computing finds natural and potentially promising applications.

§4 reviews 51 papers on *elementary combinatorial problems*: typical subjects from *computer science* like finding the maximum, merging and sorting in §4.1, and problems from *algorithmic graph theory* like finding connected components, spanning trees and shortest paths in §4.2. In each case, the papers are grouped together according to the type of machinery involved, such as general parallel computers with a shared memory and specific fixed interconnection networks.

§5 finally discusses parallelism in *combinatorial optimization.* We have been able to find 18 papers on linear programming, maximum flow, knapsack, traveling salesman and scheduling problems and on dynamic programming and branch-and-bound methods. The formidable power of parallel computing in conjunction with the firm roots of combinatorial optimization in the theory of design and analysis of algorithms and computational complexity seems to hold great promise for a further development of this area in the very near future.

We are grateful to E.L. Lawler, J. van Leeuwen, F. Maffioli and G.L. Nemhauser, who brought many papers to our attention.

1. Models

1.1. *Classification and surveys*

M.J. Flynn (1966). Very high-speed computing systems. *Proc. IEEE 54,* 1901-1909.

Four classes of parallel computers are introduced:

(1) SISD: *single instruction stream—single data stream*; one instruction is performed at a time, on one set of data; e.g., $a+b$.
(2) SIMD: *single instruction stream—multiple data stream*; one type of instruction is performed at a time, possibly on different data; e.g., $a+b$ and $c+d$.
(3) MISD: *multiple instruction stream—single data stream*; different instructions on the same data can be performed at a time; e.g., $a+b$ and $a-b$.
(4) MIMD: *multiple instruction stream—multiple data stream*; different instructions on different data can be performed at a time; e.g., $a+b$ and $c-d$.

Beyond Flynn's classification scheme, it makes sense to subdivide the last class into *synchronized* machines, which wait for each other after each set of instructions and then perform the next set, and *asynchronous* machines, which run independently and wait only if information from other processors is needed. *Systolic* algorithms are highly synchronized processes: the processing elements act rhythmically on regular streams of data passing through the network. *Distributed* algorithms are typical asynchronous processes: the processors perform their own local algorithms and communicate by sending messages every now and then.

J.T. Schwartz (1980). Ultracomputers. *ACM Trans. Programming Languages and Systems 2*, 484-521.

Distinction is made between *paracomputers*, where the processors have simultaneous access to a *shared memory*, and *ultracomputers*, where each processor communicates directly with a fixed number of other processors through an *interconnection network*. Whereas paracomputers are primarily of theoretical interest, ultracomputers are more realistic and can be quite efficient at the same time.

Important in this bibliography (although not dealt with by Schwartz) is the way in which shared memory computers handle *read* and *write conflicts*, which occur when several processors try to read from or to write into the same memory location at the same time. If read [write] conflicts are (dis-)allowed, we speak of (*no*) *simultaneous reads* [*writes*].

G. Ausiello, P. Bertolazzi (1982). Parallel computer models: an introduction. *IBM Symp. Parallel Processing*, Rome, March 1982.

In this introduction to models for parallel computation, both theoretical and practical models are considered.

L.S. Haynes, R.L. Lau, D.P. Siewiorek, D.W. Mizell (1982). A survey of highly parallel computing. *IEEE Comput. 15.1*, 9-24.

A survey of the different types of practical parallel computer structures is given.

L.G. Valiant (1983). Parallel computation. J.W. de Bakker, J. van Leeuwen (eds.). *Foundations of Computer Science IV, Distributed Systems: Part 1, Algorithms and Complexity*, Mathematical Centre Tract 158, Centre for

Mathematics and Computer Science, Amsterdam, 35-48.

This review discusses characteristics of problems that make them amenable to fast parallel computation, as well as realistic computer architectures that are suitable for such computations.

U. Vishkin (1983). *Synchronous Parallel Computation - a Survey,* Preprint, Courant Institute, New York University.

A survey of theoretical models for parallel computation (for which existing algorithms are reviewed) and of their relation to realistic machines.

1.2. *Interconnection networks*

S.H. Unger (1958). A computer oriented toward spatial problems. *Proc. IRE 46,* 1744-1750.

Introduction of the two-dimensional *mesh connected* network. Each processor is identified with an ordered pair (i,j) $(i,j = 1,...,n)$ and processor (i,j) is connected to processors $(i \pm 1,j)$ and $(i,j \pm 1)$, provided they exist.

J.S. Squire, S.M. Palais (1963). Programming and design considerations of a highly parallel computer. *Proc. AFIPS Spring Joint Computer Conf. 23,* 395-400.

Description of the *cube connected* network. It can be seen as a hypercube with processors at the vertices and interconnections along the edges.

H.S. Stone (1971). Parallel processing with the perfect shuffle. *IEEE Trans. Comput. C-20,* 153-161.

A network with interconnections that imitate a *perfect shuffle* of a deck of cards.

J.L. Bentley, H.T. Kung (1979). A tree machine for searching problems. *Proc. 1979 Internat. Conf. Parallel Processing,* 257-266.

The interconnection pattern consists of two *binary trees* with common leaves.

F.P. Preparata, J. Vuillemin (1981). The cube-connected cycles: a versatile network for parallel computation. *Comm. ACM 24,* 300-309.

The *cube connected cycles* network can be seen as a cube connected network with each processor replaced by a cyclicly connected series of processors. Each of them is connected to at most three others.

H.J. Siegel (1977). Analysis techniques for SIMD machine interconnection networks and the effects of processor address masks. *IEEE Trans Comput. C-26,* 153-161.

H.J. Siegel (1979). A model of SIMD machines and a comparison of various interconnection networks. *IEEE Trans. Comput. C-28,* 907-917.

Both papers deal with the comparison of interconnection networks. Techniques for simulating one network by another are given.

Z. Galil, W.J. Paul (1983). An efficient general-purpose parallel computer. *J. Assoc. Comput. Mach. 30,* 360-387.

A universal parallel computer, which can simulate any reasonable parallel machine efficiently.

F.P. Preparata (1982). Algorithm design and VLSI architectures. *IBM Symp. Parallel Processing,* Rome, March 1982.

Outline of desirable features for VLSI implementable networks. Some specific interconnection networks are considered in detail.

2. Complexity

2.1. *Surveys*

S.A. Cook (1981). Towards a complexity theory of synchronous parallel computation. *Enseign. Math. (2) 27,* 99-124.

This expository paper surveys machine models and complexity classes for parallel computations.

D.S. Johnson (1983). The NP-completeness column: an ongoing guide; seventh edition. *J. Algorithms 4,* 189-203.

Section 2 of this edition is a brief review of the complexity theory of parallel computing.

A *parallel RAM* with an unbounded number of processors, shared memory, simultaneous reads and no simultaneous writes is introduced, for which the *parallel computation thesis* (see §2.3) holds: the class of languages it can recognize in polynomial *time* is precisely $\mathcal{P}$SPACE, the class of languages recognizable by a sequential machine in polynomial *space*. If only a polynomial number of processors is allowed, the class of languages recognizable in parallel polynomial time shrinks from $\mathcal{P}$SPACE to $\mathcal{P}$, the class of languages recognizable in sequential polynomial time.

Many problems can be solved in *polylog parallel time,* i.e., time that is polynomially bounded in the logarithm of problem size (with unbounded parallelism); see §§3−5 for examples. By the parallel computation thesis, these problems would form the class POLYLOGSPACE of problems solvable in polylog sequential space. Other problems have been shown to be *log space complete for* $\mathcal{P}$, i.e., (i) they belong to $\mathcal{P}$ and (ii) each problem in $\mathcal{P}$ can be reduced to any of them by a transformation using logarithmic work space; see §2.2 and §5.2 for examples. If any such problem can be solved in polylog space, then $\mathcal{P} \subseteq$ POLYLOGSPACE. Since this inclusion is not expected to be true, such problems are unlikely to be solvable in polylog space or in polylog parallel time.

New classes arise if *simultaneous resource bounds* (see §2.4) are imposed.

E.g., $\mathcal{NC}$ is the class of problems solvable in polylog parallel time using only a polynomial number of processors, and $\mathcal{SC}$ is the class of problems solvable in polynomial sequential time using polylog space. Research is oriented towards questions like $\mathcal{NC} = ?\mathcal{SC}$.

2.2. *Log space completeness for $\mathcal{P}$*

S.A. Cook (1974). An observation on time-storage trade off. *J. Comput. System Sci. 9,* 308-316.

A *path system* is a quadruple $\mathcal{S} = \langle X,R,S,U \rangle$, where X is a finite set of nodes, R is a three place incidence relation on X, $S \subset X$ is a set of source nodes, and $U \subset X$ is a set of terminal nodes. $\mathcal{S}$ is *solvable* if at least one node in S is contained in the least set A such that $U \subset A$ and such that, if $y,z \in A$ and $R(x,y,z)$ holds, then $x \in A$. Cook shows that each language of time complexity $T(n)$ is reducible in storage $\log(T(n))$ to the set of strings coding solvable path systems.

N.D. Jones, W.T. Laaser (1977). Complete problems for deterministic polynomial time. *Theoret. Comput. Sci. 3,* 105-117.

The *unit resolution problem* is the problem of determining whether a propositional formula in conjunctive normal form can be proved unsatisfiable by, roughly speaking, substituting the literals in unit clauses. This problem is shown to be log space complete for $\mathcal{P}$. Corollaries give similar results for other problems. For an application, see [Dobkin, Lipton & Reiss 1979] (§ 5.2(b)).

R.E. Ladner (1975). The circuit value problem is log space complete for $\mathcal{P}$. *SIGACT News 7.1,* 18-20.

The *circuit value* problem is the problem of determining the output of a circuit consisting of AND and NOT gates, given its input. This problem is shown to be log space complete for $\mathcal{P}$ by simulating Turing machines by combinatorial circuits.

L.M. Goldschlager (1977). The monotone and planar circuit value problems are log space complete for P. *SIGACT News 9.2,* 25-29.

A circuit is *monotone* if it consists of AND and OR gates; it is *planar* if it has a cross free planar embedding. The monotone and planar circuit value problems are shown to be log space complete for $\mathcal{P}$ by a log space transformation from the circuit value problem (see [Ladner 1975] above). For an application, see [Goldschlager, Shaw & Staples 1982] (§ 5.2(a)).

2.3. *Parallel time versus sequential space*

A.K. Chandra, D.C. Kozen, L.J. Stockmeyer (1981). Alternation. *J. Assoc. Comput. Mach. 28,* 114-133.

L.M. Goldschlager (1982). A universal connection pattern for parallel computers. *J. Assoc. Comput. Mach. 29,* 1073-1086.

Statement of a hypothesis known as the *parallel computation thesis*: *time bounded parallel machines are polynomially related to space bounded sequential machines*; that is, for any function $T(n)$, the class of languages recognizable by a machine with unbounded parallelism in time $T(n)^{O(1)}$ (i.e., polynomial in $T(n)$) is equal to the class of languages recognizable by a sequential machine in space $T(n)^{O(1)}$. Evidence is given by proving the thesis for some well-behaved time bounds $T(n)$ on several parallel machine models.

J. Hartmanis, J. Simon (1974). On the power of multiplication in random access machines. *Proc. 15th Annual ACM Symp. Switching and Automata Theory,* 13-23.
V.R. Pratt, L.J. Stockmeyer (1976). A characterization of the power of vector machines. *J. Comput. System Sci. 12,* 198-221.
W.J. Savitch, M.J. Stimson (1979). Time bounded random access machines with parallel processing. *J. Assoc. Comput. Mach. 26,* 103-118.
S. Fortune, J. Wyllie (1978). Parallelism in random access machines. *Proc. 10th Annual ACM Symp. Theory of Computing,* 114-118.
A. Borodin (1977). On relating time and space to size and depth. *SIAM J. Comput. 6,* 733-744.
J.H. Reif (1982). On the power of probabilistic choice in synchronous parallel computations. M. Nielsen, E.M. Schmidt (eds.). *Proc. 9th Internat. Coll. Automata, Languages and Programming,* Lecture Notes in Computer Science 140, Springer, Berlin, 442-450.

These papers further support the parallel computation thesis.

2.4. *Simultaneous resource bounds*

N. Pippenger (1979). On simultaneous resource bounds (preliminary version). *Proc. 20th Annual IEEE Symp. Foundations of Computer Science,* 307-311.
W.L. Ruzzo (1981). On uniform circuit complexity. *J. Comput. System Sci. 22,* 365-383.
P.W. Dymond, S.A. Cook (1980). Hardware complexity and parallel computation (preliminary version). *Proc. 21st Annual IEEE Symp. Foundation of Computer Science,* 360-372.
J.W. Hong (1980). On similarity and duality of computation (extended abstract). *Proc. 21st Annual IEEE Symp. Foundations of Computer Science,* 348-359.

These papers investigate an extended version of the parallel computation thesis, formulated as follows in [Dymond & Cook 1980]: (i) *parallel time and hardware requirements are simultaneously polynomially related to sequential (Turing machine) reversal and space requirements*; (ii) *parallel time and space requirements are polynomially related.*

3. Numerical Problems

3.1. *Evaluation of expressions and recurrence relations*

R. Brent, D. Kuck, K. Maruyama (1973). The parallel evaluation of arithmetic expressions without division. *IEEE Trans. Comput. C-22,* 532-534.
D. Kuck, Y. Muraoka (1974). Bounds on the parallel evaluation of arithmetic expressions using associativity and commutativity. *Acta Inform. 3,* 203-216.
R.P. Brent (1973). The parallel evaluation of arithmetic expressions in logarithmic time. J.F. Traub (ed.). *Complexity of Sequential and Parallel Numerical Algorithms,* Academic Press, New York, 83-102.
R.P. Brent (1974). The parallel evaluation of general arithmetic expressions. *J. Assoc. Comput. Mach. 21,* 201-206.
D.J. Kuck, K. Maruyama (1975). Time bounds on the parallel evaluation of arithmetic expressions. *SIAM J. Comput. 4,* 147-162.
D.E. Muller, F.P. Preparata (1976). Restructuring of arithmetic expressions for parallel evaluation. *J. Assoc. Comput. Mach. 23,* 534-543.
S. Winograd (1975). On the parallel evaluation of certain arithmetic expressions. *J. Assoc. Comput. Mach. 22,* 477-492.
I. Munro, M. Paterson (1973). Optimal algorithms for parallel polynomial evaluation. *J. Comput. System Sci. 7,* 189-198.
K. Maruyama (1973). On the parallel evaluation of polynomials. *IEEE Trans. Comput. C-22,* 2-5.
L. Hyafil (1979). On the parallel evaluation of multivariate polynomials. *SIAM J. Comput. 8,* 120-123.
L.G. Valiant (1980). Computing multivariate polynomials in parallel. *Inform. Process. Lett. 11,* 44-45.
L.G. Valiant, S. Skyum (1981). Fast parallel computation of polynomials using few processors. J. Gruska, M. Chytil (eds.). *Mathematical Foundations of Computer Science 1981,* Lecture Notes in Computer Science 118, Springer, Berlin, 132-139.

These twelve papers deal with the parallel evaluation of arithmetic expressions. The results differ with respect to the types of expressions considered (e.g., expressions with or without division, polynomials) and the transformations allowed (using associativity, commutativity, etc.). There is also a distinction between bounded and unbounded parallelism.

Of general importance is a lemma from [Brent 1974]: *if a computation can be performed in time t with q operations and sufficiently many processors that perform arithmetic operations in unit time, then it can be performed in time $t+(q-t)/p$ with p such processors.*

P.M. Kogge, H.S. Stone (1973). A parallel algorithm for the efficient solution of a general class of recurrence equations. *IEEE Trans. Comput. C-22,* 786-793.
P.M. Kogge (1974). Parallel solution of recurrence problems. *IBM J. Res.*

Develop. 18, 138-148.
S.-C. Chen, D.J. Kuck (1975). Time and parallel processor bounds for linear recurrence systems. *IEEE Trans. Comput. C-24,* 701-717.
H.T. Kung (1976). New algorithms and lower bounds for the parallel evaluation of certain rational expressions and recurrences. *J. Assoc. Comput. Mach. 23,* 252-261.
L. Hyafil, H.T. Kung (1977). The complexity of parallel evaluation of linear recurrences. *J. Assoc. Comput. Mach. 24,* 513-521.
D.D. Gajski (1981). An algorithm for solving linear recurrence systems on parallel and pipelined machines. *IEEE Trans. Comput. C-30,* 190-206.
A.C. Greenberg, R.E. Ladner, M.S. Paterson, Z. Galil (1982). Efficient parallel algorithms for linear recurrence computation. *Inform. Process. Lett. 15,* 31-35.

These seven papers outline the results obtained on solving recurrence relations. Several types of such relations are attacked succesfully, although for the first-order recurrence problem p processors can achieve a speedup of at most $(2p+1)/3$ [Hyafil & Kung 1977].

3.2. *Numerical analysis and algebra*

D. Heller (1978). A survey of parallel algorithms in numerical linear algebra. *SIAM Rev. 20,* 740-777.

A survey of parallel techniques for problems in numerical linear algebra, such as the solution of systems of linear equations and the computation of eigenvalues, covering the literature up to 1977.

W.M. Gentleman (1978). Some complexity results for matrix computations on parallel processors. *J. Assoc. Comput. Mach. 25,* 112-115.
M.A. Franklin (1978). Parallel solution of ordinary differential equations. *IEEE Trans. Comput. C-27,* 413-420.
J.M. Lemme, J.R. Rice (1979). Speedup in parallel algorithms for adaptive quadrature. *J. Assoc. Comput. Mach. 26,* 65-71.
C.R. Jesshope (1980). The implementation of fast radix 2 transforms on array processors. *IEEE Trans. Comput. C-29,* 20-27.
O. Wing, J.W. Huang (1980). A computation model of parallel solution of linear equations. *IEEE Trans. Comput. C-29,* 632-638.
J.A.G. Jess, H.G.M. Kees (1982). A data stucture for parallel L/U decomposition. *IEEE Trans. Comput. C-31,* 231-239.
A. Borodin, J. Von Zur Gathen, J. Hopcroft (1982). Fast parallel matrix and GCD computations. *Inform. and Control 52,* 241-256.
D.J. Evans, R.C. Dunbar (1983). The parallel solution of triangular systems of equations. *IEEE Trans. Comput. C-32,* 201-204.
C.P. Arnold, M.I. Parr, M.B. Dewe (1983). An efficient parallel algorithm for the solution of large sparse linear matrix equations. *IEEE Trans. Comput. C-32,* 265-272.

J. Von Zur Gathen (1983). Parallel algorithms for algebraic problems. *Proc. 15th Annual ACM Symp. Theory of Computing,* 17-23.

More recent publications on a wide variety of problems in this very lively research area, which is, however, not of immediate interest to the theory of combinatorial optimization.

3.3. *Nonlinear optimization*

J.J. McKeown (1980). Aspects of parallel computation in numerical optimization. F. Archetti, M. Cugiani (eds.). *Numerical Techniques for Stochastic Systems,* North-Holland, Amsterdam, 297-327.

Global optimization algorithms are adapted for SIMD and MIMD computers. Parallelization is only considered at a high level, e.g. concerning the number of parallel function evaluations and local optimizations.

M.A. Franklin, N.L. Soong (1981). One-dimensional optimization on multiprocessor systems. *IEEE Trans. Comput. C-30,* 61-66.

The trade-off between two strategies for optimizing one-dimensional functions on MIMD systems is analyzed. The first strategy evaluates the function in parallel, the second one computes several function values at a time.

L.C.W. Dixon, K.D. Patel (1982). The place of parallel computation in numerical optimisation; VI parallel algorithms for nonlinear optimisation. *IBM Symp. Parallel Processing,* Rome, March 1982.

The modified Newton algorithm for nonlinear programming is parallelized, and results of an implementation on the ICL/DAP SIMD computer are presented.

4. Combinatorics

4.1. *Sorting and related problems*

(a) *sorting networks*

K.E. Batcher (1968). Sorting networks and their applications. *Proc. AFIPS Spring Joint Computer Conf. 32,* 307-314.

Networks are presented that sort n keys in $O(\log^2 n)$ time using $O(n \log^2 n)$ comparison elements. They are based on the principle of iterated merging. One network uses bitonic sequences, the other merges two ordered lists by first merging the odd and even numbered keys from both lists separately and then comparing the results.

D.E. Muller, F.P. Preparata (1975). Bounds to complexities of networks for sorting and for switching. *J. Assoc. Comput. Mach. 22,* 195-201.

A network with $O(n^2)$ elements for sorting n numbers in $O(\log n)$ time, based on enumeration sort.

M. Ajtai, J. Komlós, E. Szemerédi (1983). An $O(n \log n)$ sorting network. *Proc. 15th Annual ACM Symp. Theory of Computing*, 1-9.

A network with only $O(n \log n)$ comparison elements for sorting n numbers in $O(\log n)$ time. The basis is a network with $O(n)$ elements that splits the set of numbers in a lower and an upper half in constant time with only few errors.

(b) *shared memory computers: merging*

F. Gavril (1975). Merging with parallel processors. *Comm. ACM 18*, 588-591.

Merging two ordered sets using a small number of processors. The algorithm first splits the sets in an appropriate way and then merges the smaller parts in parallel. The processors merge the subsets sequentially.

R.H. Barlow, D.J. Evans, J. Shanehchi (1981). A parallel merging algorithm. *Inform. Process. Lett. 13*, 103-106.

Merging k sorted lists using $p \leqslant k$ processors. From one list $k-1$ elements are chosen and their place in the other lists is determined. The processors then merge the sublists obtained sequentially. The behavior of the algorithm strongly depends on the input.

(c) *shared memory computers: sorting*

S. Even (1974). Parallelism in tape-sorting. *Comm. ACM 17*, 202-204.

Synchronized MIMD.

Sorting algorithms using merge sort. They have an optimial speedup as long as the number of processors is small relative to input size.

S. Todd (1978). Algorithm and hardware for a merge sort using multiple processors. *IBM J. Res. Develop. 22*, 509-517.

Synchronized MIMD.

A parallel version of the straight merge sort algorithm. It runs in $O(n)$ time using $\log n$ processors and can be implemented in hardware.

L.G. Valiant (1975). Parallelism in comparison problems. *SIAM J. Comput. 4*, 348-355.

Valiant explores the parallelism in problems like finding the maximum, merging and sorting. If only comparisons are counted and the overhead is neglected and if the input size n is not less than the number p of processors, speedups of $\Omega(p / \log \log p)$ can be achieved. For example, $n/2$ processors can sort n keys in $O(\log n \log \log n)$ steps.

D.S. Hirschberg (1978). Fast parallel sorting algorithms. *Comm. ACM 21*,

657-661.

SIMD, shared memory, simultaneous reads, no similtaneous writes.

An algorithm to sort n keys in $O(k \log n)$ time using $n^{1+1/k}$ processors. It employs the result from [Gavril 1975] (see §4.1(b)) and a parallel bucket sort routine.

F.P. Preparata (1978). New parallel-sorting schemes. *IEEE Trans. Comput. C-27,* 669-673.

SIMD, shared memory, (no) simultaneous reads, no simultaneous writes.

Two algorithms are described to sort n numbers with enumeration sort. The first one allows read conflicts, uses the merging scheme from [Valiant 1975] (see above) and runs in $O(\log n)$ time on $n \log n$ processors, disregarding some of the overheads. The second one disallows read conflicts, uses the odd-even merging scheme from [Batcher 1968] (see §4.1(a)) and runs in $O(k \log n)$ time on $n^{1+1/k}$ processors.

Y. Shiloach, U. Vishkin (1981). Finding the maximum, merging, and sorting in a parallel computation model. *J. Algorithms 2,* 88-102.

Synchronized MIMD, shared memory, simultaneous reads, simultaneous writes provided the same value is written.

The maximum finding algorithm from [Valiant 1975] (see above) is implemented so as to achieve the same time bound while counting the overheads. Further, a merge sort algorithm is given, free of write conflicts and having the same time and processor complexities as those from [Hirschberg 1978] and [Preparata 1978] (see above).

R. Reischuk (1981). A fast probabilistic parallel sorting algorithm. *Proc. 22nd Annual IEEE Symp. Foundations of Computer Science,* 212-219.

Synchronized MIMD, shared memory, simultaneous reads, no simultaneous writes.

An algorithm to sort n keys in $O(\log n)$ average time using n processors. The set of keys is partitioned into $\lfloor \sqrt{n} \rfloor + 1$ groups, which have size $O(\sqrt{n} \log n)$ with probability close to 1, and next the groups are sorted separately.

A. Borodin, J.E. Hopcroft (1982). Routing, merging and sorting on parallel models of computation; extended abstract. *Proc. 14th Annual ACM Symp. Theory of Computing,* 338-344.

In fixed connection networks with indegrees d, oblivious routing strategies require $\Omega(\sqrt{n} / d^{3/2})$ time. On a synchronized MIMD computer with shared memory, simultaneous reads but no simultaneous writes, the merging and sorting schemes from [Valiant 1975] (see above) are implemented such that the running time is of the same order as the number of comparison steps.

C.P. Kruskal (1982). Results in parallel searching, merging and sorting (summary). *Proc. 1982 Internat. Conf. Parallel Processing,* 196-198.

Synchronized MIMD, shared memory, simultaneous reads, no simultaneous writes.

Improvements on the results from [Valiant 1975] (see above). E.g., a merge sort algorithm is given that sorts n keys in $O(\log n \log\log n / \log\log\log n)$ time using n processors.

M. Aigner (1982). Parallel complexity of sorting problems. *J. Algorithms 3,* 79-88.

Synchronized MIMD, no simultaneous reads, no simultaneous writes.

Lower and upper bounds are given on the number of comparison steps needed for selection, merging and sorting.

(d) *shared memory computers: convex hull*

D. Nath, S.N. Maheshwari, P.C.P. Bhatt (1981). Parallel algorithms for the convex hull problem in two dimensions. W. Händler (ed.). *CONPAR 81,* Lecture Notes in Computer Science 111, Springer, Berlin, 358-372.

SIMD, shared memory, (no) simultaneous reads, no simultaneous writes.

If read conflicts are allowed, the convex hull of n points in the plane can be found in $O((n/p)\log n + \log p \log n)$ time using $p \leqslant n$ processors and in $O(k \log n)$ time using $n^{1+1/k}$ processors ($k \leqslant \log n$). If read conflicts are disallowed, the same bounds still hold. After an initial sort of the points on one of the coordinates, the algorithms use a divide and conquer strategy.

(e) *mesh connected networks: permuting and sorting*

S.E. Orcutt (1976). Implementation of permutation functions in Illiac IV-type computers. *IEEE Trans. Comput. C-25,* 929-936.

SIMD, $n \times n$ mesh connected network.

This implementation of the bitonic sort from [Batcher 1968] (see §4.1(a)) performs a permutation of the n^2 elements in $O(n \log n)$ time.

C.D. Thompson, H.T. Kung (1977). Sorting on a mesh-connected computer. *Comm. ACM 20,* 263-271.

SIMD, $n \times n$ mesh connected network.

Sorting n^2 elements in snake-like (or shuffled) row-major order in $O(n)$ time, based on the odd-even (bitonic) sort from [Batcher 1968] (see §4.1(a)).

D. Nassimi, S. Sahni (1979). Bitonic sort on a mesh-connected parallel computer. *IEEE Trans. Comput. C-28,* 2-7.

SIMD, $n \times n$ mesh connected network.

Sorting n^2 elements in row-major order in $O(n)$ time by an adaptation of the bitonic sort from [Batcher 1968] (see §4.1(a)) different than that from [Thompson & Kung 1977] (see above).

D. Nassimi, S. Sahni (1980). An optimal routing algorithm for mesh-connected parallel computers. *J. Assoc. Comput. Mach. 27,* 6-29.

SIMD, k-dimensional mesh connected network ($k \geqslant 2$).

An algorithm for permuting data, which is optimal in the sense that it uses the minimum number of unit distance routing steps.

M. Kumar, D.S. Hirschberg (1983). An efficient implementation of Batcher's odd-even merge algorithm and its application in parallel sorting schemes. *IEEE Trans. Comput. C-32,* 254-264.

SIMD, $n \times n$ mesh connected network.

Another algorithm for sorting n^2 elements in $O(n)$ time based on [Batcher 1968] (see §4.1(a)).

H.-W. Lang, M. Schimmler, H. Schmeck, H. Schröder (1983). A fast sorting algorithm for VLSI. J. Díaz (ed.). *Proc. 10th Internat. Coll. Automata, Languages and Programming,* Lecture Notes in Computer Science 154, Springer, Berlin, 408-419.

Synchronized MIMD, $n \times n$ mesh connected network.

An algorithm for sorting n^2 elements in $O(n)$ time, based on odd-even transposition sort. A systolic version is presented that runs in $O(n)$ time using $O(n^2)$ cells.

(f) *other interconnection networks: permuting and sorting*

G. Baudet, D. Stevenson (1978). Optimal sorting algortihms for parallel computers. *IEEE Trans. Comput. C-27,* 84-87.

SIMD, (i) linearly connected network, (ii) mesh connected network, (iii) perfect shuffle network.

As long as the number of processors stays small compared to the number of keys, odd-even transposition sort has an optimal speedup on (i) and the methods from [Batcher 1968] (see §4.1(a)) on (ii) and (iii).

D. Nassimi, S. Sahni (1982). Parallel permutation and sorting algorithms and a new generalized connection network. *J. Assoc. Comput. Mach. 29,* 642-667.

SIMD, (i) cube connected network, (ii) perfect shuffle network.

Sorting algorithms similar to that from [Preparata 1978] (see §4.1(c)) are given for networks (i) and (ii); sorting n elements requires $O(k \log n)$ time using $n^{1+1/k}$ processors ($k \leqslant \log n$). Further, permutation algorithms that are faster by a constant factor are given for these machines.

L.G. Valiant (1982). A scheme for fast parallel communication. *SIAM J. Comput. 11,* 350-361.

(Synchronized) MIMD, cube connected network.

Description of a randomized two-phase algorithm that performs permutations on an n-node cube connected network in $O(\log n)$ time with probability close to 1. In the first phase each packet is sent to a randomly chosen node, in the second phase the packets find their way to their destination.

J.H. Reif, L.G. Valiant (1983). A logarithmic time sort for linear size networks. *Proc. 15th Annual ACM Symp. Theory of Computing,* 10-16.

(Synchronized) MIMD, cube connected cycles network.

A randomized algorithm for sorting n keys on an n-node cube connected cycles network in $O(\alpha \log n)$ time with probability at least $1-n^{\alpha}$, using ideas from [Valiant 1982] (see above).

(g) *interconnection networks: data transmission*

D. Nassimi, S. Sahni (1981). Data broadcasting in SIMD computers. *IEEE Trans. Comput. C-30,* 101-106.

SIMD, (i) mesh connected network, (ii) cube connected network, (iii) perfect shuffle network.

Algorithms for data transmission, with particular attention for read and write conflicts.

L.G. Valiant, G.J. Brebner (1981). Universal schemes for parallel communication. *Proc. 13th Annual ACM Symp. Theory of Computing,* 263-277.

Description of a randomized data transmission algorithm based on ideas from [Valiant 1982] (see §4.1(f)). With probability close to 1, it runs in $O(\log n)$ time on networks like the n-cube.

4.2. *Graph theory*

Many parallel algorithms have been developed for problems on graphs, such as finding connected components, transitive closures, spanning trees and shortest paths. Throughout this subsection, graphs (digraphs) have n vertices and m edges (arcs).

(a) *shared memory computers*

E. Reghbati (Arjomandi), D.G. Corneil (1978). Parallel computations in graph theory. *SIAM J. Comput. 7,* 230-237.

SIMD, shared memory, simultaneous reads, no simultaneous writes.

Finding the connected components of a graph and the weakly and strongly connected components of a digraph has the same time complexity as finding the transitive closure and therefore requires $O(\log^2 n)$ time using n^3 processors. Of three different bounded parallel graph search techniques, namely k-depth, breadth-depth and breadth-first search, the last one achieves a bound close to optimal if $m = \Theta(n^2)$.

D.M. Eckstein, D.A. Alton (1977). *Parallel Searching of Non-Sparse Graphs,*

Technical report 77-02, Department of Computer Science, University of Iowa, Iowa City.

Synchronized MIMD, shared memory, simultaneous reads, no simultaneous writes.

Depth-first search and breadth-first search of a graph can be performed in $O(n+m/p)$ time using p processors. These algorithms are essentially optimal if $n = O(m)$.

D.S. Hirschberg, A.K. Chandra, D.V. Sarwate (1979). Computing connected components on parallel computers. *Comm. ACM 22,* 461-464.

SIMD, shared memory, simultaneous reads, no simultaneous writes.

The connected components of a graph and hence the transitive closure of an $n \times n$ symmetric Boolean matrix can be obtained in $O(\log^2 n)$ time using n^2 processors and even using $n \lceil n/\log n \rceil$ processors. The connected components are built up by merging smaller parts together.

C. Savage, J. Ja'Ja' (1981). Fast, efficient parallel algorithms for some graph problems. *SIAM J. Comput. 10,* 682-691.

SIMD, shared memory, simultaneous reads, no simultaneous writes.

Algorithms are presented for finding the connected and biconnected components, the bridges and a minimum spanning tree of a graph in $O(\log^2 n)$ time. The number of processors used is small enough to make the parallel implementations relatively efficient.

D. Nath, S.N. Maheshwari (1982). Parallel algorithms for the connected components and minimal spanning tree problems. *Inform. Process. Lett. 14,* 7-11.

SIMD, (i) shared memory without simultaneous reads and simultaneous writes, (ii) perfect shuffle network, (iii) orthogonal trees network.

The algorithm from [Hirschberg, Chandra & Sarwate 1979] (see above) is modified such that it finds the connected components of a graph or a minimum spanning tree in a weighted connected graph in $O(\log^2 n)$ time using n^2 ($n \lceil n/\log n \rceil$) processors on a shared memory model *without* read (and write) conflicts. Implementations on networks (ii) and (iii) require $O(\log^2 n \log\log n)$ time.

F.Y. Chin, J. Lam, I-N. Chen (1982). Efficient parallel algorithms for some graph problems. *Comm. ACM 25,* 659-665.

SIMD, shared memory, simultaneous reads, no simultaneous writes.

The algorithm from [Hirschberg, Chandra & Sarwate 1979] (see above) is modified to run in $O(n^2/p + \log^2 n)$ time using p processors ($p \geqslant n$), i.e., in $O(\log^2 n)$ time for $p = n \lceil n/\log^2 n \rceil$. Slight adaptations of the algorithm find the weakly connected components of a digraph, a spanning forest and a minimum spanning tree.

Y. Shiloach, U. Vishkin (1982). An $O(\log n)$ parallel connectivity algorithm.

J. Algorithms 3, 57-67.

Synchronized MIMD, shared memory, simultaneous reads, simultaneous writes (one (unknown) processor succeeds).

In this strong model, the connected components of a graph can be found in only $O(\log n)$ time using $n+2m$ processors.

J. Ja'Ja', J. Simon (1982). Parallel algorithms in graph theory: planarity testing. *SIAM J. Comput. 11,* 314-328.

Synchronized MIMD, shared memory, simultaneous reads, no simultaneous writes.

Finding the connected components and planarity testing can be done in $O(\log^2 n)$ time using a polynomial number of processors.

L. Kučera (1982). Parallel computation and conflicts in memory access. *Inform. Process. Lett. 14,* 93-96.

SIMD, shared memory, simultaneous reads, simultaneous writes.

In models where simultaneous writes are allowed under certain conditions, the connected components of a graph can be found in only $O(\log n)$ time using a polynomial number of processors. Similar results hold for other graph problems such as finding a minimum spanning tree.

N. Deo, C.Y. Pang, R.E. Lord (1980). Two parallel algorithms for shortest path problems. *Proc. 1980 Internat. Conf. Parallel Processing,* 244-253.

MIMD, shared memory, simultaneous reads, no simultaneous writes.

The Moore/Pape-D'Esopo algorithm for finding all shortest paths from one vertex and the Floyd-Warshall algorithm for finding the shortest paths between all pairs of vertices are implemented on the Heterogeneous Element Processor, an MIMD machine.

N. Deo, Y.B. Yoo (1981). Parallel algorithms for the minimum spanning tree problem; summary. *Proc. 1981 Internat. Conf. Parallel Processing,* 188-189.

MIMD, shared memory, simultaneous reads, no simultaneous writes.

Three minimum spanning tree algorithms are considered: Prim-Dijkstra, requiring $O(n^2/p+np)$ time on p processors $(p \leqslant n)$; Kruskal, achieving no speedup; and Sollin, requiring $O((n^2/p)\log n)$ time on p processors $(p \leqslant n)$.

J.A. Wisniewski, A.H. Sameh (1982). Parallel algorithms for network routing problems and recurrences. *SIAM J. Algebraic Discrete Meth. 3,* 379-394.

The single source shortest path problem can be stated as solving systems of the form $x = Ax+b$ in the regular algebra of Carré. Solution methods, mostly known from linear algebra, are parallelized.

(b) *interconnection networks*

K.N. Levitt, W.H. Kautz (1972). Cellular arrays for the solution of graph problems. *Comm. ACM 15,* 789-801.

Synchronized MIMD, two-dimensional mesh connected network.

The Floyd-Warshall shortest path algorithm, Kruskal's minimum spanning tree algorithm and other graph theoretical algorithms are implemented on special purpose hardware, buildable using (V)LSI technology.

L.J. Guibas, H.T. Kung, C.D. Thompson (1979). Direct VLSI implementation of combinatorial algorithms. *Caltech Conf. VLSI,* 509-525.

Synchronized MIMD, $n \times n$ mesh connected network.

For the transitive closure problem and a special class of dynamic programming problems, algorithms are designed that run in $O(n)$ time. They are suitable for VLSI implementation.

F.L. Van Scoy (1980). The parallel recognition of classes of graphs. *IEEE Trans. Comput. C-29,* 563-570.

Synchronized MIMD, $n \times n$ mesh connected network.

On this network, Warshall's transitive closure algorithm is implemented to run in $O(n)$ time.

D. Nassimi, S. Sahni (1980). Finding connected components and connected ones on a mesh-connected parallel computer. *SIAM J. Comput. 9,* 744-757.

SIMD, k-dimensional mesh connected network.

Consider a graph with $n = l^k$ vertices, none of which has degree more than d. Following [Hirschberg, Chandra & Sarwate 1979] (see §4.2(a)), the connected components can be found in $O(k^3(k+d)l \log l)$ time on a k-dimensional mesh connected network with l^k processing elements. For $d = 2$, the algorithm is modified to run in $O(k^4 l)$ time. The connected ones problem, a connected connectivity problem, requires $O(k^6 l)$ time.

M.J. Atallah, S.R. Kosaraju (1982). Graph problems on a mesh-connected processor array (preliminary version). *Proc. 14th Annual ACM Symp. Theory of Computing,* 345-353.

SIMD, $n \times n$ mesh connected network.

On this network, the bridges, the articulation points, the length of a shortest cycle and a minimum spanning tree of a graph are found in $O(n)$ time.

S.E. Hambrusch (1983). VLSI algorithms for the connected component problem. *SIAM J. Comput. 12,* 354-365.

Synchronized MIMD, k-dimensional mesh connected network.

Algorithms for finding connected components in $O(n^{1+1/k})$ time on a k-dimensional mesh connected network with n processors, suitable for VLSI implementation.

J.L. Bentley (1980). A parallel algorithm for constructing minimum spanning trees. *J. Algorithms 1,* 51-59.

Synchronized MIMD, tree structured network.

A parallel version of the Prim-Dijkstra minimum spanning tree algorithm, requiring $O(n \log n)$ time on a tree structured machine consisting of $O(n/\log n)$ processors.

E. Dekel, D. Nassimi, S. Sahni (1981). Parallel matrix and graph algorithms. *SIAM J. Comput. 10,* 657-675.

SIMD, (i) perfect shuffle network, (ii) cube connected network.

On both networks, two $n \times n$ matrices can be multiplied in $O(n/m + \log n)$ time using mn^2 processors and in $O(n^2/m + m(n/m)^{2.61})$ time using m^2 processors $(1 \leq m \leq n)$. These algorithms are applied to solve several problems on graphs, e.g., shortest path problems.

(c) *distributed algorithms*

R.G. Gallager, P.A. Humblet (1979). *Minimum Weight Spanning Trees*, Technical report LIDS-P-906, Massachusetts Institute of Technology, Cambridge.

MIMD, full interconnection.

This distributed minimum spanning tree algorithm is an asynchronous implementation of Sollin's method and requires $O(n \log n)$ time using n processors.

P.A. Humblet (1981). *A Distributed Algorithm for Minimum Weight Directed Spanning Trees*, Technical report LIDS-P-1149, Massachusetts Institute of Technology, Cambridge.

MIMD, full interconnection.

This algorithm for finding n minimum spanning arborescences, one rooted at each vertex, parallelizes the Chu-Liu/Edmonds/Bock algorithm and requires $O(n^2)$ time using n processors.

K.M. Chandy, J. Misra (1982). Distributed computation on graphs: shortest path algorithms. *Comm. ACM 25,* 833-837.

MIMD, full interconnection.

A distributed version of Ford's algorithm for finding all shortest paths from a single vertex, terminating properly if negative cycles occur.

5. Combinatorial Optimization

5.1. *Well-solvable problems: polylog parallel algorithms*

(a) *sequencing and scheduling*

The machine scheduling problems that have been subjected to parallelization will be indicated below by means of the concise notation of Graham, Lawler, Lenstra & Rinnooy Kan (*Ann. Discrete Math. 5* (1979), 287-326).

E. Dekel, S. Sahni (1983A). Binary trees and parallel scheduling algorithms. *IEEE Trans. Comput. C-32,* 307-315.

SIMD, shared memory, no simultaneous reads, no simultaneous writes.

Binary trees turn out to be useful for all sorts of parallel computations. E.g., the partial sums of a series of n numbers can be computed in $O(\log n)$ time using $O(n/\log n)$ processors

scheduling problems considered	*sequential time*	*parallel time*	*#processors*
$P\|p_j=1,r_j\|L_{max}$	$O(n \log n)$	$O(\log^2 n)$	$O(n)$
$1\|pmtn,r_j\|L_{max}$	$O(n \log n)$	$O(\log^2 n)$	$O(n)$
$1\|prec,p_j=1,r_j\|L_{max}$	$O(n^2)$	$O(\log^2 n)$	$O(n^2/\log n)$
$1\|prec,pmtn,r_j\|L_{max}$	$O(n^2)$	$O(\log^2 n)$	$O(n^2/\log n)$
$1\|\|\Sigma U_j$	$O(n \log n)$	$O(\log^2 n)$	$O(n)$
$1\|p_j=1\|\Sigma w_j U_j$	$O(n \log n)$	$O(\log^2 n)$	$O(n)$

E. Dekel, S. Sahni (1981). *A Parallel Matching Algorithm for Convex Bipartite Graphs and Applications to Scheduling,* Technical report 81-3, Computer Science Department, University of Minnesota, Minneapolis.

E. Dekel, S. Sahni (1982). A parallel matching algorithm for convex bipartite graphs. *Proc. 1982 Internat. Conf. Parallel Processing,* 178-184.

SIMD, shared memory, no simultaneous reads, no simultaneous writes.

A bipartite graph with vertex sets $V = \{v_1,...,v_n\}$, $W = \{w_1,...,w_m\}$ and edge set E is *convex* if $\forall v_j \in V \; \exists l(j), u(j)$: $(v_j,w_i) \in E \Leftrightarrow l(j) \leqslant i \leqslant u(j)$. The binary tree method provides the basis of an algorithm for finding a maximum matching in such a graph in $O(\log^2 n)$ time using $O(n)$ processors.

scheduling problems considered	*sequential time*	*parallel time*	*#processors*
$1\|p_j=1,r_j\|f_{max}$	$O(n^2 \log n)$	$O(\log^3 n)$	$O(n^2/\log^2 n)$
$1\|p_j=1,r_j\|\Sigma w_j U_j$	$O(n^2)$	$O(\log^2 n)$	$O(n^2/\log n)$
$P2\|pmtn,p_j=1,prec\|C_{max}$	$O(n^2)$	$O(\log^2 n)$	$O(n^3/\log n)$

E. Dekel, S. Sahni (1983B). Parallel scheduling algorithms. *Oper. Res. 31,* 24-49.

SIMD, shared memory, no simultaneous reads, no simultaneous writes.

The developed algorithms rely on parallel sorting and the parallel computation of partial sums.

scheduling problems considered	*sequential time*	*parallel time*	*#processors*
$P\|pmtn\|C_{max}$	$O(n)$	$O(\log n)$	$O(n/\log n)$
$Q\|\|\Sigma C_j$	$O(n \log mn)$	$O(\log mn)$	$O(m^2 n^2)$
$1\|p_j=1\|\Sigma U_j$	$O(n)$	$O(\log n)$	$O(n^2)$
$1\|p_j=1\|\Sigma w_j U_j$	$O(n \log n)$	$O(\log n)$	$O(n^2)$ [1]
$1\|r_j\|\max_j\{\max\{e(E_j),f(T_j)\}\}$	$O(n \log n)$	$O(\log n)$	$O(n^2)$ [2]
$P\|r_j,C_j=r_j+p_j\|m$	$O(n \log n)$	$O(\log n)$	$O(n^2)$ [3]

[1] See [Dekel & Sahni 1983A] above for a different algorithm.

[2] $E_j = \max\{0,r_j-(C_j-p_j)\}$ is the *earliness* of job j; e & f are nondecreasing functions with $e(0) = f(0) = 0$.

[3] The *channel assignment problem*: minimize the number of identical parallel

machines needed to process a set of jobs with fixed starting times.

(b) *miscellaneous*

N. Megiddo (1983). Applying parallel computation algorithms in the design of serial algorithms. *J. Assoc. Comput. Mach. 30,* 852-865.

The efficiency of serial algorithms for one problem may be improved by exploiting the parallelism in other problems. E.g., Valiant's and Preparata's parallel sorting algorithms (see §4.1(c)) turn out to be useful for cost-effective resource allocation and parallel all-pairs shortest path algorithms for the minimum ratio cycle problem. Other examples are given for scheduling and spanning tree problems.

5.2. *Well-solvable problems, log space complete for* $\mathcal{P}$

(a) *maximum flow*

L.M. Goldschlager, R.A. Shaw, J. Staples (1982). The maximum flow problem is log space complete for P. *Theoret. Comput. Sci. 21,* 105-111.

The result stated in the title is obtained through a log space transformation from the monotone circuit value problem (see [Goldschlager 1977], §2.2).

Y. Shiloach, U. Vishkin (1982). An $O(n^2 \log n)$ parallel MAX-FLOW algorithm. *J. Algorithms 3,* 128-146.

Synchronized MIMD, shared memory, simultaneous reads, simultaneous writes provided the same value is written.

A p-processor system is developed that solves the maximum flow problem on an n-vertex network in $O(n^3(\log n)/p)$ time, for $p \leq n$. The algorithm is closely related to the sequential methods due to Dinic and Karzanov, that use layered networks.

D.B. Johnson, S.M. Venkatesan (1982). Parallel algorithms for minimum cuts and maximum flows in planar networks (preliminary version). *Proc. 23rd Annual IEEE Symp. Foundations of Computer Science,* 244-254.

Synchronized MIMD, shared memory, simultaneous reads, no simultaneous writes.

Computing the maximum flow in planar directed n-vertex networks requires $O(\log^3 n)$ time using $O(n^4)$ processors and $O(\log^2 n)$ time using $O(n^6)$ processors; in planar undirected networks, $O(\log^2 n)$ time and $O(n^4)$ processors suffice. The results are based on the fact that the minimum cut capacity in a network N is equal to the length of a minimum forward cut cycle in a network related to the dual network of N.

(b) *linear programming*

D. Dobkin, R.J. Lipton, S. Reiss (1979). Linear programming is log-space hard for *P. Inform. Process. Lett. 8*, 96-97.

In conjunction with Khachian's algorithm, this implies that linear programming is log space complete for $\mathcal{P}$. The log space transformation starts from the unit resolution problem (see [Jones & Laaser 1974], §2.2).

B. Kamdoum (1982). Speeding up the primal simplex algorithm on parallel computer. *SIGMAP Newsletter 31, 19-23.*

Each pivot step of the simplex method can be executed p times faster when p processors are available and p is small compared to the number n of variables and the number m of constraints.

N. Megiddo (1982). *Poly-log Parallel Algorithms for LP with an Application to Exploding Flying Objects,* Unpublished manuscript.

Megiddo has previously shown that linear programs can be solved by an $O(m)$ sequential algorithm when n is fixed *(J. Assoc. Comput. Mach.* (to appear)). A parallel implementation of this method runs in $O(\log^n m)$ time. Improvements lead to an algorithm requiring $O(\log^{n-1} m \log\log m)$ time and a probabilistic algorithm requiring $O(\log m (\log\log m)^{n-2})$ expected time on a parallel RAM model. An interesting application to warfare is presented.

5.3. *$\mathcal{NP}$-hard problems and enumerative methods*

(a) knapsack

A.C.-C. Yao (1982). On parallel computation for the knapsack problem. *J. Assoc. Comput. Mach. 29*, 898-903.

In parallel computation models with real arithmetic, solution of the knapsack problem with n real inputs requires an exponential number of processors if a running time of at most $\sqrt{n}/2$ is to be achieved.

(b) *traveling salesman*

E.A. Pruul (1975). *Parallel Processing and a Branch-and-Bound Algorithm,* M.Sc. thesis, Cornell University, Ithaca, NY.

MIMD, shared memory.

A p-processor implementation of the subtour elimination algorithm for the traveling salesman problem with n cities is developed and simulated on a sequential machine. For small p and n, the simulated parallel algorithm runs faster than the traditional serial method.

M.J. Quinn, N. Deo (1983). *A Parallel Approximate Algorithm for the Euclidean Traveling Salesman Problem,* Report CS-83-105, Computer Science

Department, Washington State University, Pullman.

MIMD, shared memory, simultaneous reads, no simultaneous writes.

The farthest-insertion heuristic for the Euclidean traveling salesman problem is implemented to run on the Heterogeneous Element Processor (see [Deo, Pang & Lord 1980], §4.2(a)) with p processors in $O(n^2/p + np)$ time.

(c) *dynamic programming*

J. Casti, M. Richardson, R. Larson (1973). Dynamic programming and parallel computers. *J. Optim. Theory Appl. 12,* 423-438.

Finite-stage dynamic programming procedures allow a natural parallelization. E.g., at each stage, the various states can be dealt with simultaneously by different processors.

D. Al-Dabass (1980). Two methods for the solution of the dynamic programming algorithm on a multiprocessor cluster. *Optimal Control Appl. Methods 1,* 227-238.

The efficieny of the algorithms developed in the previous paper is analyzed on a master-slave architecture.

P. Bertolazzi, M. Pirozzi (undated). *Parallel Algorithms for Dynamic Programming Algorithms,* Unpublished manuscript.

After a review of the methods proposed in the above two papers, for two classes of problems an implementation on a special configuration is shown to reduce computational complexity.

(d) *branch-and-bound*

O.I. El-Dessouki, W.H. Huen (1980). Distributed enumeration on between computers. *IEEE Trans. Comput. C-29,* 818-825.

MIMD, full interconection.

Note. In the title, read 'network' for 'between'.

A distributed branch-and-bound algorithm. Each processor determines by itself which part of the tree it searches for an optimal solution.

F.W. Burton, G.P. McKeown, V.J. Rayward-Smith, M.R. Sleep (1982). Parallel processing and combinatorial optimization. L.B. Wilson, C.S. Edwards, V.J. Rayward-Smith (eds.). *Combinatorial Optimization III,* University of Stirling, 19-36.

MIMD, r-ary n-cube.

Distributed branch-and-bound algorithms are considered for execution on the r-ary n-cube, a processor network that can be built using VLSI techniques. Each processor is invoked by one of its neighbors.

9

Location and Network Design

R.T. Wong
Purdue University, West Lafayette

CONTENTS

Selecting the optimal network configuration plays a vital role in many applications such as transportation, distribution and computer communication systems. Two important areas of network configuration research are:

(1) *location theory* which concerns the optimal placement of facilities such as warehouses, bank accounts, or data files;

(2) *network design* which encompasses a broad range of configuration questions including the optimal layout of a transportation network and the best method of connecting a set of terminals to a central computer.

Many location and network design problems are most naturally cast as discrete choice models thus giving rise to *combinatorial optimization problems*. Indeed, the fields of combinatorial optimization and network configuration have enjoyed a close and fruitful relationship. Historically, the study of location algorithms has been a fertile area for the development and application of combinatorial optimization concepts and techniques. Network design research is also beginning to benefit from recent advances in combinatorial optimization methods.

The purpose of this annotated bibliography is to provide a guide to recent (from 1981 to date) research on location and network design problems. In order to facilitate our presentation we first give a brief description and classification of the basic problems in location and network design.

The simplest location model is the *uncapacitated plant location problem*:

$$\text{minimize} \quad \sum_{i=1}^{m} \sum_{j=1}^{n} c_{ij} x_{ij} + \sum_{j=1}^{n} f_j y_j \tag{1}$$

$$\text{subject to} \quad \sum_{j=1}^{n} x_{ij} = 1 \qquad (1 \leqslant i \leqslant m), \tag{2}$$

$$x_{ij} \leqslant y_j \qquad (1 \leqslant i \leqslant m, 1 \leqslant j \leqslant n), \tag{3}$$

$$x_{ij} = 0 \text{ or } 1, \; y_j = 0 \text{ or } 1 \qquad (1 \leqslant i \leqslant m, 1 \leqslant j \leqslant n). \tag{4}$$

There are m customers and n potential facility locations. Variable $x_{ij} = 1$ if customer i is served at location j and $x_{ij} = 0$ otherwise. Constraint (2) requires each customer to be served by a facility and c_{ij} represents the cost of servicing customer i at location j. Variable y_j is 1 or 0 depending on whether or not a facility is placed at location j and f_j is the fixed cost of using location j. Constraint (3) states that customers cannot be served at location j unless a facility is placed there.

The *capacitated plant location problem* is described by (1)−(4) with an additional constraint limiting the total customer load a facility may service.

Another related model is the *generalized plant location problem* which is an uncapacitated plant location problem where the number of facilities used is limited by the constraint

$$\sum_{j=1}^{n} y_j = p \tag{5}$$

and we wish to maximize instead of minimize the objective function.

Special versions of the above models arise when the location problem is defined on a network. In this case, the nodes correspond to both customer sites and potential facility locations and each cost c_{ij} is the shortest path distance in the network between nodes i and j. Also we will let n denote the number of nodes in the network.

For example, the model given by (1)−(5) defined on a network with all $f_j = 0$ is the classical *p-median problem* (Hakimi, *Oper. Res. 12* (1964), 450-459).

The *vertex constrained p-center problem* is given by (2)−(5) defined on a network where the objective is to minimize the maximum distance between any customer (node) and its closest facility.

If we allow facilities to be located anywhere on the network (on arcs as well as nodes) the problem becomes the classical *p-center problem*. If we regard any point on the network as a customer site we obtain the *continuous p-center problem*.

Another class of network location problems occurs when customer sites and facilities are restricted to nodes and each facility can only service customers within a specified travel distance. If a customer is not served a penalty cost is incurred. The *minimum cost partial covering problem* is to locate a set of facilities to minimize the sum of the cost of establishing facilities and the penalty cost for not serving customers. If we have an additional constraint limiting the total number of facilities used the model becomes the *minimum cost maximal covering problem*.

For our *general network design models* each arc has a length (or per unit flow cost) and a fixed charge associated with using the arc. The *fixed charge design problem* is to find the subset of arcs which minimizes the weighted sum of shortest paths between nodes plus the sum of fixed charges incurred. A related model is the *budget design problem* which is the same as the fixed charge design problem except that the objective is to minimize the weighted sum of shortest paths subject to an additional budget constraint limiting the total sum of fixed charges incurred.

For certain applications we wish to find a network design that corresponds to a tree subnetwork. Each arc has an associated cost (or fixed charge). The *Steiner tree problem on a graph* is to find an (undirected/directed) subtree whose total arc cost is minimal and which connects a designated set of nodes to a given root node.

For the *capacitated spanning tree problem* we have a specific traffic load associated with each node and a capacity limit for each arc. Our objective is to find a minimum cost spanning tree connecting a given root node to the other nodes. The design tree must satisfy the constraint that the total traffic load passing through an arc to the root node cannot exceed the arc capacity.

The above set of location and network design models is quite broad and allows us to cover a wide spectrum of the current research on network configuration. However, we surely cannot cover all of the interesting topics in this area. Also we have concentrated on recent (from 1981 to date) location and network design research. For information about earlier research efforts, the reader may refer to a variety of survey papers and books given in this annotated bibliography.

1. Location Theory: Books and Surveys

1.1. *Books*

R.L. Francis, J.A. White (1974). *Facility Layout and Location: an Analytical*

Approach, Prentice-Hall, Englewood Cliffs, NJ.

This book is the first one to focus on location theory. It contains a highly recommended discussion of non-network location problems especially location models on the plane which we do not discuss.

G.Y. Handler, P.B. Mirchandani (1979). *Location on Networks,* The MIT Press, Cambridge, MA.

An excellent introduction to location theory which discusses most of the location models used in this bibliography. Most of the references are from 1977 or earlier.

P.B. Mirchandani, R.L. Francis (eds.) (to appear). *Discrete Location Theory,* Wiley, New York.

An advanced text which covers most of the major topics in location theory. It includes chapters by Berman, Chiu, Larson, Odoni; Cornuéjols, Nemhauser, Wolsey; Erlenkotter; Francis, Lowe, Tansel; Goldman; Hakimi; Kolen and Tamir; Magnanti and Wong; Burkard; Handler; Hansen, Thisse and Wendall; Krarup and Pruzan; Mirchandani.

J.F. Thisse, H.G. Zoller (eds.) (1983). *Locational Analysis of Public Facilities,* North-Holland, Amsterdam.

This collection of papers has contributions both to location theory and the economics of public facility location decisions. Several of the chapters including ones by Hansen, Peeters and Thisse, Halpern and Maimon, and Wolsey discuss topics related to the models used in this bibliography.

N. Christofides (1975). *Graph Theory: an Algorithmic Approach,* Academic Press, New York.

Two of its chapters discuss the p-median and p-center problems on a network.

E. Minieka (1978). *Optimization Algorithms for Networks and Graphs,* Marcel Dekker, New York.

One chapter concerns network location problems with the primary emphasis on the 1-center and 1-median problems.

R. C. Larson, A.R. Odoni (1981). *Urban Operations Research,* Prentice-Hall, Englewood Cliffs, NJ.

A very nice discussion of the p-median and p-center problems is given in one of its chapters.

1.2. *Surveys*

R.L. Francis, L.F. McGinnis, J.A. White (1983). Locational analysis. *European J. Oper. Res. 12,* 220-252.

An overview of planar location, warehouse layout, network location and plant location problems is given. The emphasis is on thoroughly tested models which can be solved efficiently.

B.C. Tansel, R.L. Francis, T.J. Lowe (1983). Location on networks: a survey; part I: the p-center and p-median problems. *Management Sci. 29,* 482-497.
B.C. Tansel, R.L. Francis, T.J. Lowe (1983). Location on networks: a survey; part II: exploiting tree network structure. *Management Sci. 29,* 498-511.

This excellent overview concentrates on tree network location research over the period 1978 to 1981.

L.F. McGinnis (1977). A survey of recent results for a class of facilities location problems. *AIIE Trans. 9,* 11-18.

This review of capacitated plant location algorithms describes the state of the art around 1975.

J. Krarup, P. Pruzan (1979). Selected families of location problems. *Ann. Discrete Math. 5,* 327-387.

This survey presents a comprehensive discussion of the p-median and p-center problems for both planar location and network location models.

J. Krarup, P. Pruzan (1983). The simple plant location problem: survey and synthesis. *European J. Oper. Res. 12,* 36-81.

This very useful and complete review covers uncapacitated and capacitated plant location problems. Formulations, algorithms and interesting properties of these models are discussed.

L.A. Wolsey (1983). Fundamental properties of certain discrete location problems. J.F. Thisse, H.G. Zoller (eds.). *Locational Analysis of Public Facilities,* North-Holland, Amsterdam, 331-355.

Some of the recent work on generalized plant location problems is surveyed in order to help explain recent computational successes. The primary emphasis is on primal and dual approximation algorithms.

J. Halpern, O. Maimon (1982). Algorithms for the m-center problems: a survey. *European J. Oper. Res. 10,* 90-99.

A unified framework is presented and used to relate and compare various p-center network location algorithms. The survey shows that a number of p-center algorithms can be viewed as solving a series of suitably defined covering problems.

2. Uncapacitated Plant Location

2.1. *Optimization algorithms: bounding procedures*

Schrage (*Math. Programming Stud. 4* (1975), 118-132) obtained good computational results by using a special compact inverse version of the simplex method to solve the linear programming relaxation of (1)−(4). This relaxation frequently has integer optimal solutions and yields excellent lower bounds for branch and bound routines. One difficulty with this approach is that linear programs with variable upper bound constraints of the form $x_{ij} \leqslant y_j$ are frequently massively degenerate.

M.J. Todd (1982). An implementation of the simplex method for linear programming problems with variable upper bounds. *Math. Programming 23,* 34-49.

An alternative to Schrage's method is given which circumvents the massive degeneracy inherent in the variable upper bound constraints and which can be implemented using triangular basis factorizations.

N.I. Yanev (1981). Solution of a simple allocation problem. *U.S.S.R. Comput. Math. and Math. Phys. 21,* 95-103.

A result is given which shows that the plant location problem can be solved by applying the simplex method to the linear programming relaxation of (1)−(4) and restricting all basic feasible solutions to be integer valued. This result has already been given by Guignard and Spielberg (*Ann. Discrete Math. 1* (1977), 247-271). Also Yanev's modification of the simplex method to exploit this result is not correct.

T.L. Magnanti, R.T. Wong (1981). Accelerating Benders decomposition: algorithmic enhancement and model selection criteria. *Oper. Res. 29,* 464-484.

General methods for selecting good Benders cuts and for evaluating alternative problem formulations are given. When specialized to plant location type problems, the cut generation technique leads to very efficient algorithms that exploit the underlying structure of these models.

2.2. *Valid inequalities and facets*

For several important discrete optimization problems such as the traveling salesman problem, the best available algorithms use information about the facial structure of the convex hull of the feasible solutions. Thus, there has been significant interest in deriving facets for the plant location problem.

G. Cornuejols, M.L. Fisher, G.L. Nemhauser (1977). On the uncapacitated location problem. *Ann. Discrete Math. 1,* 163-177.

Besides summarizing the worst-case error bounds for the greedy heuristic and linear programming relaxations of the generalized plant location problem, this paper gives a characterization of the fractional extreme points of the linear programming relaxation (1)−(4). A class of inequalities to eliminate some of these fractional solutions is given.

M. Guignard (1980). Fractional vertices, cuts and facets of the simple plant location problem. *Math. Programming Stud. 12,* 150-162.

The above fractional extreme point characterization is used to describe a related set of valid inequalities. Some of these cuts are shown to be plant location facets.

G. Cornuejols, J.M. Thizy (1982). Some facets of the simple plant location polytope. *Math. Programming 23,* 50-74.
D.C. Cho, E.L. Johnson, M.W. Padberg, M.R. Rao (1983). On the uncapacitated plant location problem. I: valid inequalities and facets. *Math. Oper. Res. 8,* 579-589.
D.C. Cho, M.W. Padberg, M.R. Rao (1983). On the uncapacitated plant location problem II: facets and lifting theorems. *Math. Oper. Res. 8,* 590-612.

These three noteworthy papers characterize and describe various classes of plant location facets. Their methods involve using vertex packing facets to derive plant location inequalities and the sequential lifting of valid cuts to obtain facets.

2.3. *Approximation algorithms*

G. Cornuejols, J.M. Thizy (1982). A primal approach to the simple plant location problem. *SIAM J. Algebraic Discrete Meth. 3,* 504-510.

A Lagrangean relaxation approach applied to the *dual* of the LP relaxation of (1)−(4) is combined with a greedy-interchange procedure to produce a very accurate heuristic.

D.S. Hochbaum (1982). Heuristics for the fixed cost median problem. *Math. Programming 22,* 148-162.

An $O(n^2m)$ heuristic is proposed that has a worst-case error ratio of $\ln(n)$. The procedure utilizes Chvátal's (*Math. Oper. Res. 4* (1979), 233-235) worst-case analysis of a greedy set covering heuristic.

2.4. *Special case: tree networks*

A. Kolen (1983). Solving covering problems and the uncapacited plant location problem on trees. *European J. Oper. Res. 12,* 266-278.

An $O(n^3)$ algorithm is given for the plant location problem on a tree network.

2.5. *Related models: dynamic facility location*

D. Erlenkotter (1981). A comparative study of approaches to dynamic location problems. *European J. Oper. Res. 6,* 133-143.

The computational performances of seven heuristic procedures are compared and evaluated.

T.J. Van Roy, D. Erlenkotter (1982). A dual-based procedure for dynamic facility location. *Management Sci. 28,* 1091-1105.

A very efficient branch and bound algorithm is presented which uses a dual ascent routine to generate lower bounds. The procedure can be extended to models that include price-sensitive demands, concave facility costs and capacitated facilities.

2.6. *Related models: multi-commodity location*

For this extension, each customer has demands for every commodity. Every opened facility must designate exactly one type of commodity that it will supply. The multi-commodity plant location problem is to find the optimal set of facilities such that each customer's demand for every commodity is satisfied.

A.W. Neebe, B.M. Khumawala (1981). An improved algorithm for the multi-commodity location problem. *J. Oper. Res. Soc. 32,* 143-149.

Khumawala's uncapacitated plant location branch and bound algorithm is extended to the multi-commodity case.

J. Karkazis, T.B. Boffey (1981). The multi-commodity facilities location problem. *J. Oper. Res. Soc. 32,* 803-814.

The computational performance of a dual ascent approach and a Lagrangean dual-based algorithm are compared.

3. Capacitated Plant Location

3.1. *Optimization algorithms*

T.J. Van Roy (to appear). A cross decomposition algorithm for capacitated facility location. *Oper. Res.*

T.J. Van Roy (1981). Cross decomposition for mixed integer programming with applications to facility location. J.P. Brans (ed.). *Operational Research '81,* North-Holland, Amsterdam, 579-587.

A computationally effective algorithm based on cross decomposition (Van Roy, *Math. Programming 25* (1983), 46-63) is presented. The approach incorportates attractive features of both Lagrangean relaxation and Benders decomposition.

N. Christofides, J.E. Beasley (1983). Extensions to a Lagrangean relaxation approach for the capacitated warehouse location problem. *European J. Oper. Res. 12,* 19-28.

A computationally effective branch and bound procedure based upon Lagrangean relaxation and problem reduction (variable fixing) tests is presented.

E. Bartezzaghi, A. Colorni, P.C. Palermo (1981). A search tree algorithm for plant location problems. *European J. Oper. Res. 7,* 371-379.

An implicit enumeration scheme based upon a suitable ranking of all possible combinations of plant locations is presented along with computational results.

G.R. Bitran, V. Chandru, D.E. Sempolinski, J.F. Shapiro (1981). Inverse optimization: an application to the capacitated plant location problem. *Management Sci. 27,* 1120-1141.

Inverse optimization concerns the fact that the solution for a Lagrangean subproblem derived from an integer program is optimal for a related integer program with a suitably adjusted right hand side vector. A parametric solution technique based on inverse optimization is investigated for the capacitated plant location problem.

B.M. Baker (1982). Linear relaxations of the capacitated warehouse location problem. *J. Oper. Res. Soc. 33,* 475-479.

A new type of valid inequality is derived. Unfortunately, this cut is shown to be implied by the constraints of a standard LP relaxation.

3.2. *Approximation algorithms*

S.K. Jacobsen (1983). Heuristics for the capacitated plant location model. *European J. Oper. Res. 12,* 253-261.

A number of uncapacitated plant location heuristics (including add, drop and various exchange procedures) are adapted and evaluated for the capacitated case.

3.3. *Related models: location with transshipment nodes*

The capacitated plant location model is defined on a network which includes transshipment nodes that can only be used if a facility is located at the node.

M. Guignard (1982). *Preprocessing and Optimization in Network Flow Problems with Fixed Charges,* Report 47, Department of Statistics, University of Pennsylvania, Philadelphia.

Preprocessing techniques that include constraint disaggregation, generation of implied precedent and capacity constraints, and coefficient reduction are described to improve the performance of optimization algorithms.

M. Guignard (1982). *Large Duality Gaps in Network Location Problems,* Report 48, Department of Statistics, University of Pennsylvania, Philadelphia.
Systematic techniques for strengthening the linear programming relaxation of this location model are given.

4. Generalized Plant Location

G. Cornuejols, G.L. Nemhauser, L.A. Wolsey (1980). A canonical representation of simple plant location problems and its applications. *SIAM J. Algebraic Discrete Meth. 1,* 261-272.
Several new formulations for the generalized plant location problem are given. Also certain classes of solution algorithms are shown to require almost complete enumeration of the feasible solution set.

J. Krarup, P. Pruzan (1981). *Assessment of Approximate Algorithms: the Error Measure's Crucial Role*, Report 81-7, Institute of Datalogy, University of Copenhagen.
The worst-case error measure of Cornuejols, Fisher and Nemhauser (*Management Sci. 23* (1977), 789-810) for generalized location models is discussed and comments are made about how appropriate it is.

L.A. Wolsey (1982). Maximizing real-valued submodular functions: primal and dual heuristics for location problems. *Math. Oper. Res. 7,* 410-425.
When the constraint limiting the total number of facilities is replaced by a knapsack constraint, an adapted greedy heuristic always attains 35% of the optimal (maximization) problem value. Bounds are also given for primal and dual greedy heuristics applied to an LP relaxation of the problem.

G.L. Nemhauser, L.A. Wolsey (1981). Maximizing submodular set functions: formulations and analysis of algorithms. *Ann. Discrete Math. 11,* 279-301.
Submodularity results are used to guide a branch and bound approach to the generalized location model. Worst-case error bounds are given for when the branch and bound search is prematurely terminated.

5. Location on a Network: p-Medians

5.1. *Optimization algorithms*

R.D. Galvao (1981). A note on Garfinkel, Neebe and Rao's LP decomposition for the p-median problem. *Transportation Sci. 15,* 175-182.
Some additional computational experience with a Dantzig-Wolfe decomposition approach originally proposed by Garfinkel *et al.* (*Transportation Sci. 8* (1974), 217-236) is presented along with a disussion of LP degeneracy.

R.D. Galvao (1981). A graph theoretical bound for the p-median problem. *European J. Oper. Res. 6,* 162-165.
L. Morgenstern (1983). A note on Galvao's 'A graph theoretical bound for the p-median problem'. *European J. Oper. Res. 12,* 404-405.
The first paper, with a correction and extension given in the second paper, discusses lower bounds produced by minimum spanning tree calculations on the underlying network.

N. Christofides, J.E. Beasley (1982). A tree search algorithm for the p-median problem. *European J. Oper. Res. 10,* 196-204.
P.B. Mirchandani, A. Oudjit, R.T. Wong (1981, revised 1983). *'Multidimensional' Extensions and a Nested Dual Approach for the m-Median Problem,* Working paper, Rensselaer Polytechnic Institute, Troy, NY.
Both papers use a similar Langrangean relaxation approach which utilizes a very efficient plant location algorithm due to Erlenkotter (*Oper. Res. 26* (1978), 992-1009) as a subroutine. The first reference also discusses penalties and problem reduction (variable fixing) tests for branch and bound approaches while the second gives a special dual simplex routine for updating dual variables.

5.2. *Approximation algorithms*

This section deals with the class of p-median problems where the nodes are randomly distributed points over the unit square and c_{ij}, the cost of servicing node i at node j, is the euclidean distance between the points representing nodes i and j. Let n be the number of nodes and $p(n)$ be the number of medians which increases as a function of n.

M.L. Fisher, D.S. Hochbaum (1980). Probabilistic analysis of the planar K-median problem. *Math. Oper. Res. 5,* 27-34.
If $p(n)/n < (\log n)/n$ then a special aggregation heuristic is proved to be asymptotically optimal. See also Ch.6, §5.4.

C.H.Papadimitriou (1981). Worst-case and probabilistic analysis of a geometric location problem. *SIAM J. Comput. 10,* 542-557.
The Euclidean version of the p-median problem (as described above) is shown to be $\mathcal{NP}$-hard. For $(\log n)/n < p(n)/n < 1/\log n$, a partitioning heuristic is proved to be asymptotically optimal. See also Ch.6, §5.4.

M. Haimovich (in preparation). *Asymptotic Properties of Geometric Location Problems,* Doctoral dissertation, Department of Aeronautics and Astronautics, Massachusetts Institute of Technology, Cambridge.
Papadimitrou's partitioning heuristic is shown to be asymptotically optimal whenever $p(n)/n$ approaches zero as n approaches infinity. Also the partitioning heuristic is extended to solve situations with a nonuniform distribution of

customers.

5.3. *Related models: stochastic networks*

Mirchandani and Odoni (*Transportation Sci. 13* (1979), 85-97) have extended the concept of median location to stochastic networks where the travel cost on some arcs and/or the demands at some nodes may be discrete random variables.

J.R. Weaver, R.L. Church (1983). Computational procedures for location problems on stochastic networks. *Transportation Sci. 17,* 168-180.
A moderately successful Lagrangean relaxation approach is given.

Mirchandani, Oudjit and Wong (see §5.1) have shown how to transform the stochastic p-median problem to a generalized plant location model and to use regular p-median routines as a solution procedure.

6. Location on a Network: p-Centers

6.1. *Optimization algorithms: tree networks*

A.M. Farley (1982). Vertex centers of trees. *Transportation Sci. 16,* 265-280.
S.M. Hedetniemi, E.J. Cockayne, S.T. Hedetniemi (1981). Linear algorithms for finding the Jordan center and path center of a tree. *Transportation Sci. 15,* 98-114.
Both papers present $O(n)$ algorithms for the vertex constrained 1-center problem and various extensions of it.

R. Chandrasekaran, A. Daughety (1981). Location on tree networks: p-center and n-dispersion problems. *Math. Oper. Res. 6,* 50-57.
An $O(n^2 \log p)$ algorithm is given for the p-center problem. A polynomial algorithm is also given for the location of p facilities so as to maximize the minimum distance between any pair of facilities.

R. Chandrasekaran, A. Tamir (1982). Polynomially bounded algorithms for locating p-centers on a tree. *Math. Programming 22,* 304-315.
An $O(n^4 \log n)$ algorithm for the vertex constrained continuous p-center problem is given.

N. Megiddo, A. Tamir, E. Zemel, R. Chandrasekaran (1981). An $O(n \log^2 n)$ algorithm for the kth longest path in a tree with applications to location problems. *SIAM J. Comput. 10,* 328-337.
The previous result is improved with an $O(n \log^2 n)$ algorithm. Also an $O(n \cdot \min\{p \log^2 n, n \log p\})$ algorithm is given for the continuous p-center problem.

6.2. *Optimization algorithms: general networks*

E. Minieka (1981). A polynomial time algoritm for finding the absolute center of a network. *Networks 11,* 351-355.

An $O(n^3)$ algorithm is given for the 1-center problem on a general network.

G.Y. Handler, M. Rozman (1982). *The Continuous m-Center Problem on a Network,* Working paper 755, Faculty of Management, Tel Aviv University.

A relaxation procedure which uses column and row generation techniques is used to solve the continuous p-center problem.

6.3. *Approximation algorithms*

D.S. Hochbaum, D.B. Shmoys (to appear). A best possible heuristic for the k-center problem. *Math. Oper. Res.*

A polynomial algorithm with a worst-case error ratio of 2 is described for the vertex constrained p-center problem. Since Hsu and Nemhauser (*Discrete Appl. Math. 1* (1979), 209-216) have shown that finding a polynomial heuristic with a worst-case ratio < 2 would imply $\mathcal{P}=\mathcal{NP}$, the given error bound appears optimal.

6.4. *Related models*

B.C. Tansel, R.L. Francis, T.J. Lowe (1982). A biobjective multifacility minimax location problem on a tree network. *Transportation Sci. 16,* 407-429.

The p-center problem on a tree network objective function is used along with a second objective involving the maximum of the weighted distances between facilities. An algorithm for finding the efficient frontier of this biobjective problem is given.

B.C. Tansel, R.L. Francis, T.J. Lowe, M.L. Chen (1982). Duality and distance constraints for the nonlinear p-center problem and covering problem on a tree network. *Oper. Res. 30,* 725-743.

The p-center problem on a tree network is generalized to include travel costs that are nonlinear functions of distance. A polynomial solution algorithm is described.

A. Tamir, E. Zemel (1982). Locating centers on a tree with discontinuous supply and demand regions. *Math. Oper. Res. 7,* 183-197.

A polynomial algorithm is given for an extension of the p-center problem on a tree network where the set of customers and set of potential locations are arbitrarily designated sets of points on the tree network.

D.R. Shier, P.M. Dearing (1983). Optimal locations for a class of nonlinear single-facility location problems on a network. *Oper. Res. 31,* 292-303.

Necessary and sufficient conditions for characterizing local optima are discussed for single facility problems on general networks. The objective function is a weighted sum of the travel distances under the l_r norm.

7. Location on a Network: Covering

7.1. *Minimum cost partial covering*

A.W.J. Kolen (1982). *Location Problems on Trees and in the Rectilinear Plane,* Ph.D. dissertation, Centre for Mathematics and Computer Science, Amsterdam.

A.W.J. Kolen (1983). Solving covering problems and the uncapacitated plant location problem on trees. *European J. Oper. Res. 12,* 266-278.

An elegant $O(n^2)$ algorithm is given for the minimum cost partial covering problem on a tree network.

Y. Gurevich, L. Stockmeyer, U. Vishkin (1982). *Solving $\mathcal{NP}$-Hard Problems on Graphs That Are Almost Trees and an Application to Facility Location Problems,* Research report RC 9348, IBM Thomas J. Watson Research Center, Yorktown Heights, NY.

An $O(e \cdot (6r)^{\lceil k/2 \rceil})$ algorithm is given for a special case of the minimum cost partial covering problem on general networks. The total number of arcs is e; r is the number of facilities in the optimal solution and k is the number of arcs that must be removed from the network so that it becomes a tree.

A.J. Hoffman, A.W.J. Kolen, M. Sakarovich (1984). Totally balanced and greedy matrices. *SIAM J. Algebraic Discrete Math.*

The algorithm of Kolen (1983) is extended to give a polynomial algorithm for a generalized covering model involving totally balanced matrices.

7.2. *Minimum cost maximal covering*

W.L. Hsu (1982). The distance-domination number of trees. *Oper. Res. Lett. 1,* 96-100.

An $O(n^2 \log p)$ algorithm is given for a special case of the minimum cost maximal covering problem for tree networks. The same basic algorithm can also be applied to the p-median problem on tree networks.

N. Megiddo, E. Zemel, S.L. Hakimi (1983). The maximum coverage location problem. *SIAM J. Algebraic Discrete Meth. 4,* 253-261.

An $O(n^2 \log p)$ dynamic programming algorithm is given that handles the general minimum cost covering problem for tree networks.

M.W. Broin, T.J. Lowe (1983). *A Dynamic Programming Algorithm for Covering Problems with (Greedy) Totally-Balanced Constraint Matrices,* Working paper, Krannert Graduate School of Management, Purdue University, West Lafayette, IN.

The problem of the previous reference is generalized to a covering model involving totally balanced matrices. A polynomial dynamic programming solution algorithm is given.

8. Network Design: General Models

8.1. *Books and surveys*

D.E. Boyce (ed.) (1979). *Transportation Res. 13B.1.*

This special issue on transportation network design problems includes articles by Dantzig *et al.*, Abdulaal and LeBlanc, Los, Rothengatter, Boyce and Soberanes, and Pearman.

T.L. Magnanti, R.T. Wong (1984). Network design and transportation planning: models and algorithms. *Transportation Sci. 8,* 1-55.

A comprehensive synthesis of research on fixed charge network design models is given. Problems with continuous design variables (including convex and concave cost models) are extensively discussed.

R.T. Wong (to appear). Introduction to network design problems: models and algorithms. M. Florian (ed.). *Transportation Planning,* North-Holland, Amsterdam.

This tutorial paper introduces and discusses some important classes of network design problems and their solution methods.

8.2. *Optimization algorithms: bounding procedures*

G. Gallo (1981). *A New Branch and Bound Algorithm for the Network Design Problem,* Working paper L81.1, Istituto di Elaborazione dell'Informazione, Università di Pisa.

A new lower bounding method is presented for the budget design problem. The technique uses feasibility information available at vertices in the branch and bound search tree.

R.L. Rardin (1982). *Tight Relaxations of Fixed Charge Network Flow Problems,* Report J-82-3, School of Industrial and Systems Engineering, Georgia Institute of Technology, Atlanta.

An improved formulation which has a tighter linear programming relaxation is given for a capacitated version of the fixed charge design problem. Initial computational results are quite promising.

G. Gallo (1983). Lower planes for the network design problem. *Networks 13,* 411-426.

A new type of lower bounding procedure which uses information about the penalties for removing certain arcs is presented for the budget design problem.

T.L. Magnanti, P. Mireault, R.T. Wong (1983). *Tailoring Benders Decomposition for Network Design,* Working paper, Operations Research Center, Massachusetts Institute of Technology, Cambridge.

A special preprocessing technique which eliminates non-optimal variables is combined with acceleration techniques for Benders decomposition to provide an algorithm for the fixed charge design problem.

8.3. *Valid inequalities and facets*

E.L. Johnson, M. Pieri, P. Pieroni (1982). *Facets for Polyhedra of Fixed-Charge Shortest Path Problems,* Report 89, Dipartimento di ricera operativa e scienze statistiche, Università di Pisa.

A previously derived characterization of facets for multiple right hand choice linear programs is used to derive facets for the fixed charge design problem.

M.W. Padberg, T.J. Van Roy, L.A. Wolsey (1982). *Valid Linear Inequalities for Fixed Charge Problems,* CORE discussion paper 8232, Center for Operations Research and Econometrics, Louvain-la-Neuve.

Mixed zero-one integer programming problems that can be viewed as simplified versions of fixed charge design problems are studied. Some facets and methods for recognizing violated facet inequalities are given for these problems.

8.4. *Approximation algorithms*

R.T. Wong (1980). Worst-case analysis of network design problem heuristics. *SIAM J. Algebraic Discrete Meth. 1,* 51-63.

Under the assumption that $\mathcal{P} \neq \mathcal{NP}$, any polynomial time heuristic for the budget design problem with unit flow demands that always produces a feasible solution is shown to have a worst case error ratio between $n^{1-\epsilon}$ and n where ϵ is any positive constant.

R.T. Wong (1984). *Probabilistic Analysis of a Network Design Problem Heuristic,* Working paper, Krannert Graduate School of Management, Purdue University, West Lafayette, IN.

For budget design problems with a sufficiently large budget, a modified add heuristic is shown to be asymptotically optimal.

M. Los, C. Lardinois (1982). Combinatorial programming, statistical optimization and the optimal transportation network problem. *Transportation Res. 16B,* 89-124.

A branch and bound algorithm and an add-drop heuristic are given for the fixed charge design problem. A statistical analysis technique is presented for evaluating the heuristic's accuracy.

8.5. *Related models: nonlinear cost functions*

H.H. Hoang (1982). Topological optimization of networks: a nonlinear mixed integer model employing generalized Benders' decomposition. *IEEE Trans. Automat. Control AC-27,* 164-169.

Generalized Benders decomposition is used to solve a generalization of the fixed charge design problem with nonlinear flow costs.

H. Poorzahedy, M.A. Turnquist (1982). Approximate algorithms for the discrete network design problem. *Transportation Res. 16B,* 45-55.

Two heuristic procedures are presented for a variant of the budget design problem which has special nonlinear constraints and flow costs related to traffic equilibrium conditions for the network flows.

9. NETWORK DESIGN: STEINER TREE PROBLEMS ON A GRAPH

Recall that the Steiner tree problem on a graph is to find an (undirected/directed) subtree whose total arc cost is minimal and which connects a designated set of nodes S to a given root node.

9.1. *Optimization algorithms*

M.L. Shore, L.R. Foulds, P.B. Gibbons (1982). An algorithm for the Steiner problem in graphs. *Networks 12,* 323-333.

Minimal spanning tree computations are used to provide rudimentary bounds for a branch and bound routine.

J.E. Beasley (1984). An algorithm for the Steiner problem in graphs. *Networks 14,* 147-160.

Computational results with several effective Lagrangean relaxations combined with problem reduction (variable fixing) tests are presented.

R.T. Wong (1984). A dual ascent approach for Steiner tree problems on a directed graph. *Math. Programming. 28,* 271-287.

Computational results for an efficient dual ascent procedure are given. The method can be viewed as a generalization of Edmonds' directed spanning tree algorithm and the Bilde-Krarup-Erlenkotter location procedure.

9.2. *Approximation algorithms*

L. Kou, G. Markowsky, L. Berman (1981). A fast algorithm for Steiner trees.

Acta Inform. 15, 141-145.

An $O(n^3)$ heuristic based on minimal spanning tree computations is given for the undirected network case. Its worst-case error ratio is 2.

G.F. Sullivan (1982). *Approximation Algorithms for Steiner Tree Problems,* Technical report 249, Department of Computer Science, Yale University, New Haven, CT.

For every k, $0 \leqslant k \leqslant |S|-1$, an $O((n-|S|)^k |S|^2+n^3)$ heuristic for the undirected network case is described which has a worst-case error ratio of $(2-k/(|S|-1))$.

9.3. *Special cases*

J.A. Wald, C.J. Colbourn (1982). Steiner trees in outerplanar graphs. *Proc. 13th Southeastern Conf. Combinatorics, Graph Theory and Computing,* 15-22.

An $O(n)$ algorithm is given for the special case of the Steiner tree problem on outerplanar graphs.

J.A. Wald, D.J. Colbourn (1983). Steiner trees, partial 2-trees, and minimum IFI networks. *Networks 13,* 159-167.

A linear time algorithm is given for the special case of partial 2-tree (also known as series-parallel) graphs. Since outerplanar graphs are a special case of series-parallel graphs, this result improves the one given above.

10. NETWORK DESIGN: CAPACITATED SPANNING TREES

10.1. *Surveys*

R.R. Boorstyn, H. Frank (1977). Large-scale network topological optimization. *IEEE Trans. Comm. COM-25,* 29-47.

This review of various computer network design problems includes an informative discussion of capacitated spanning tree research until 1976.

H. Kobayashi (1981). Communication network design and control algorithms. J.K. Skwirzynski (ed.). *Concepts in Multi-User Communication,* Sythoff & Noordhoff, 373-406.
H. Kobayashi (1982). *Communication Network Design and Control Algorithms - a Survey,* Computer science research report RC9233, IBM Thomas J. Watson Research Center, Yorktown Heights, NY (revision of [Kobayashi 1981]).

Several sections of this survey discuss research from 1977 to 1980 on the capacitated spanning tree problem and related models.

10.2. *Optimization algorithms*

B. Gavish (1980). New algorithms for the capacitated minimal directed tree

problem. *Proc. 1980 IEEE Internat. Conf. Circuits and Computers,* Port Chester, 996-1000.
Computational tests utilizing a new integer programming problem formulation and various relaxation techniques are given.

B. Gavish (1982). Topological design of centralized computer networks. *Networks* 12, 355-377.
A Benders decomposition approach applied to the new formulation of the previous reference is shown to perform poorly. Very good results were obtained with a Langrangean approach for the related problem of finding minimal cost degree constrained spanning trees.

B. Gavish (1983). Formulations and algorithms for the capacitated minimal directed tree problem. *J. Assoc. Comput. Mach. 30,* 118-132.
An extension of the integer programming formulations of [Gavish 1980, 1982] leads to a Lagrangean relaxation approach capable of solving large-scale problems.

A. Kershenbaum, R.R. Boorstyn (1983). Centralized teleprocessing network design. *Networks 13,* 279-293.
A new lower bounding scheme based upon node partitioning is used in a branch and bound procedure capable of solving small to moderate sized problems.

10.3. *Approximation algorithms*

A. Kershenbaum, R.R. Boorstyn, R. Oppenheim (1980). Second-order greedy algorithms for centralized teleprocessing network design. *IEEE Trans. Comm. COM-28,* 1835-1838.
A new two level approximation method is defined by embedding existing heursitics inside an arc insertion/deletion routine. Empirical tests show that it is quite effective.

R.L. Sharma (1983). Design of an economical multidrop network topology with capacity constraints. *IEEE Trans. Comm. COM-31,* 590-591.
The computational performances of several previously proposed heuristics are compared.

10.4. *Related models*

A.W. Shogan (1983). Constructing a minimal-cost spanning tree subject to resource constraints and flow requirements. *Networks 13,* 169-190.
Tree design problems with resource constraints and arc capacities are solved with a Lagrangean relaxation algorithm.

10

Vehicle Routing

N. Christofides
Imperial College of Science and Technology, London

CONTENTS

The *vehicle routing problem* (VRP) is defined as follows. Given a graph G with a set of vertices V and a set of arcs A, let c_{ij} be the cost (distance, time, etc.) of arc $a_{ij} \in A$, and let $v_0 \in V$ be a distinguished vertex. Consider a depot x_0 to be located at v_0 and a set of customers x_i, $i = 1,...,n$, to be located at a subset $V' \subset V$ of the vertices of G, with q_i the demand (quantity to be delivered or collected) of customer i. A set of m vehicles exists at the depot with the capacity of the jth vehicle being Q_j. The VRP is then the problem of routing a subset of the vehicles to supply the customer demands and return back to the depot. The objective is to minimize the cost of the routes (e.g. the total distance traveled).

Various complications to the above basic VRP are introduced by practical considerations. These include:

(i) *Alternative objectives,* e.g. minimization of the number of vehicles used in the routing.

(ii) *Dynamic demand:* (a) demand depending on the *inventory level* at the customer; in this case the maximum storage capacity and the rate of consumption of the customer are given; (b) probabilistic demand.

Some practical considerations are not stated explicitly, but their implicit introduction leads to variations of the VRP. Amongst these are the following:

(iii) *Multiperiod problems* where in a given period (say a week) a certain customer must be visited more than once, the frequency of call depending on the same considerations as in (ii-a) above.

(iv) Design of *fixed routes* which can be operated unchanged over a given period even though the demmand is changing. This is a problem clearly intimately related to case (ii-b) above.

A problem that is closely related (in practical terms) to the VRP is the *arc routing problem* (ARP) defined as follows. Let G be the graph as in the case of the VRP, with the depot at v_0 and the same vehicle fleet located there. A customer in the ARP is an arc $(v_i, x_j) \equiv a_k \in A$ and the demand q_k corresponds to the quantity delivered or collected by traversing this arc. The problem is to route the vehicles so that the subset of arcs $A' \subset A$ where customers are located is traversed and the customer demand satisfied. The same objectives and generalizations apply as for the VRP.

The single-vehicle versions of the VRP and ARP are the *traveling salesman problem* (TSP) and *rural postman problem* (RPP) respectively; in case the graph induced by A' is connected, the RPP is known as the *Chinese postman problem* (CPP). These problems are not considered in the bibliography. However, a particularly simplified version of the VRP (when vehicle capacity is set to ∞ but all m vehicles are *required* to be used) is called the m-TSP and some note is made of this problem.

Problems where the commodity to be delivered (collected) does not only originate (terminate) at the depot, but can orginate at any location and be destined for any other location, also appear often in practice. The central problem in these cases is the *dial-a-ride* problem and some versions of the *school bus routing* problem. Hence, some attention is given to these two problems in the bibliography.

1. General Routing Problems: Surveys and Classification

S. Eilon, C. Watson-Gandy, N. Christofides (1971). *Distribution Management,* Griffin, London.

This is one of the first textbooks on distribution, and part II of the book is devoted to routing problems including the TSP and VRP. For the VRP the savings, 3-optimal and other heuristics are described and compared. Another chapter is devoted to the computation of expected distances of TSP and VRP solutions. Problems of vehicle loading and minimization of vehicle fleet size are also considered. An important contribution of this book is the list of test problems which have now become the standard test problems in the VRP literature.

B. Gavish, S. Graves (1979). *The TSP and Related Problems,* Working paper 7906, Graduate School of Management, University of Rochester, NY.

Gives a mixed integer programming formulation of the TSP, VRP, multidepot VRP, dial-a-ride, and school bus problems. A Benders decomposition procedure is applied to these formulations but computational results are given only for the TSP.

L. Bodin, B. Golden (1981). Classification in vehicle routing and scheduling. *Networks 11,* 97-108.

Discusses numerous variations of vehicle routing and scheduling problems and provides a taxonomic structure and a hierarchy moving from the very simple to the very complex problems. Solution strategies for solving vehicle routing problems are classified. Excellent list of references.

J.K. Lenstra, A.H.G. Rinnooy Kan (1981). Complexity of vehicle routing and scheduling problems. *Networks 11,* 221-227.

Investigates the computational complexity of a class of vehicle routing and scheduling problems and compiles the worst case performance results of heuristics for these problems.

L. Bodin, B. Golden, A. Assad, M. Ball (1983). The state of the art in the routing and scheduling of vehicles and crews. *Computers and Oper. Res. 10,* 63-212.

Gives a very detailed survey of routing and scheduling problems.

N. Christofides (1984). *Distribution Systems: Routing and Location,* Report IC-OR-84-7, Imperial College of Science and Technology, London.

This report is essentially the first draft of a manuscript to appear as a book with the same title. The routing half of the report deals with formulations, algorithms, heuristics and applications of the TSP, constrained TSP, CPP, constrained CPP, RPP, VRP, ARP and more general routing problems. The manuscript is of an advanced pedagogical and unifying nature, illustrated with examples.

2. THE BASIC VRP: SURVEYS

N. Christofides (1976). Vehicle routing. *RAIRO Rech. Opér. 10,* 55-70.

Classifies both optimization and approximation methodes for the VRP. Properties of feasible solutions are described which can be used to reduce the computational effort involved in solving VRPs. Some computational comments are included.

B. Golden (1976). *Recent Developments in Vehicle Routing,* Presented at the Bicentennial Conf. Mathematical Programming, November 1976.

Gives an excellent historical survey of the VRP, and comments on computational experiments with various heuristic algorithms. Derives formulae for calculating the number of feasible solutions to a VRP.

N. Christofides, A. Mingozzi, P. Toth (1979). The vehicle routing problem. N.

Christofides, A. Mingozzi, P. Toth, C. Sandi (eds.). *Combinatorial Optimization,* Wiley, Chichester, Ch. 11.

Gives a brief survey of the VRP together with two new heuristics. Compares the computational performance of the new heuristics with three other heuristics from the literature including the savings and sweep algorithms. Computational results are given for fourteen test problems.

T. Magnanti (1981). Combinatorial optimization and vehicle fleet planning: perspectives and prospects. *Networks 11,* 179-213.

Discusses the VRP, giving various formulations and its relationship to the TSP. Describes various algorithms for the optimal solution of these formulations and also describes optimization based heuristics. An excellent list of references is included.

C. Watson-Gandy, L. Foulds (1981). The vehicle routing problem - a survey. *New Zealand Oper. Res. 9,* 73-92.

Gives a survey of the VRP and lists 84 references.

L. Schrage (1981). Formulation and structure of more complex/realistic routing and scheduling problems. *Networks 11,* 229-232.

Classifies the features encountered in real VRPs, and discusses the modelling approaches that can represent such features.

N. Christofides (1985). Vehicle routing. E.L. Lawler, J.K. Lenstra, A.H.G. Rinnooy Kan, D.B. Shmoys (eds.). *The Traveling Salesman Problem,* Wiley, Chichester, Ch.12.

Gives a detailed survey of the formulations and corresponding algorithms for the exact solution of the VRP. Algorithms discussed include those based on set covering, Benders decomposition and state space relaxation. The paper also gives a classification and description of heuristics including the many versions of savings, sweep, and the two-phase methods of Fisher & Jaikumar and Christofides, Mingozzi & Toth. Computational comments and a long list of references are included.

3. The Basic VRP

3.1. *Savings and insertion*

G. Dantzig, J. Ramser (1959). The truck dispatching problem. *Management Sci. 6,* 80-91.

The first mention of the VRP in the literature and the problem definition.

G. Clarke, J. Wright (1964). Scheduling of vehicles from a central depot to a number of delivery points. *Oper. Res. 12,* 568-581.

Describes a heuristic that came to be known as the *savings* method for the

VRP. The heuristic consists of ordering all possible links between pairs of customers according to a criterion of *distance saved if the link is made.* The procedure then scans this ordered list examining the links (from the best to the worst) and accepting any link whose inclusion in the solution being constructed leaves the solution feasible. The paper has become a classic in the VRP literature because of its early date, and also because despite of its simplicity the heuristic produces reasonable results.

T. Gaskell (1967). Bases for vehicle fleet scheduling. *Oper. Res. Quart. 18,* 281-287.

Examines both the sequential and parallel versions of the savings algorithm and introduces alternatives to the savings measure. The resulting algorithms are tested and compared on six test problems.

K. Knowles (1967). *The Use of a Heuristic Tree-Search Algorithm for Vehicle Routing and Scheduling,* Presented at the OR Conference, Exeter, England, 1967.

Introduces a binary tree search algorithm where at each node two choices are considered: (i) picking the link with the largest saving, or (ii) picking the link with the second largest saving and ignoring the link in (i). The heuristic produces large improvements over the savings algorithm, but it is very time consuming.

P. Yellow (1970). A computational modification to the savings method of vehicle scheduling. *Oper. Res. Quart. 21,* 281-283.

Introduces a computational improvement to the savings algorithm which reduces both the memory and time requirements of the method. A 1000-customer problem was solved on an IBM360/50 in five minutes.

R. Mole, S. Jameson (1976). A sequential route-building algorithm employing a generalized savings criterion. *Oper. Res. Quart. 27,* 503-511.

The savings criterion is generalized and parameters are introduced which could lead to different shape routes depending on their value. A sequential route building algorithm using this criterion is described and tested on ten problems.

B. Golden, T. Magnanti, H. Nguyen (1977). Implementing vehicle routing algorithms. *Networks 7,* 113-148.

Efficient data structures are introduced for the implementation of the savings algorithm. Modifications and extensions to the algorithm permit the solution of VRPs with hundreds of customers in a matter of seconds. For example, a 600 customer problem was solved in 20 seconds on an IBM370/168. The savings algorithm is compared to the sweep and Tyagi's algorithms and is extended to deal with multiple-depot VRPs. These multiple-depot VRPs are of the independent depot variety and an example with 600 customers and two depots was solved in 55 seconds on the same computer quoted above.

B. Golden (1977). Evaluating a sequential vehicle routing algorithm. *AIIE Trans. 9,* 204-208.

Evaluates the worst case behavior of the sequential version of the savings heuristic for the VRP.

B. Williams (1982). Vehicle scheduling: proximity priority searching. *J. Oper. Res. Soc. 33,* 961-966.

Gives a sequential tour building heuristic based on assigning customers to routes according to a set of proximity criteria.

L. Bodin (1983). *Solving Large Vehicle Routing and Scheduling Problems in Small Core,* Working paper MS/S 83-027, University of Maryland, College Park.

Gives an efficient implementation of the savings algorithm for the VRP. This implementation has enabled 300-customer problems to be solved in 64 kbytes of memory, and 900-customer problems to be solved in 128 kbytes of memory and in less than nine seconds on an IBM3033 computer.

3.2. *Cluster first-route second*

B. Gillett, L. Miller (1974). A heuristic algorithm for the vehicle-dispatch problem. *Oper. Res. 22,* 340-349.

Describes a two-phase heuristic for the VRP that has come to be known as the *sweep.* In the first phase, customers are clustered into routes. Customers enter a given cluster in the order in which they are swept by a rotating imaginary ray emanating from the depot. When the capacity of the cluster reaches the vehicle capacity, a new cluster is started as the sweep continues. In the second phase the customers within each cluster are sequenced to produce routes. Good computational results are reported.

R. Russell (1974). *Efficient Truck Routing for Industrial Refuse Collection,* Presented at ORSA/TIMS Meeting, Puerto Rico, October 1974.

The sweep algorithm is modified to enhance its computational speed and applied to a real-life refuse collection operation. Some implementation aspects are discussed.

P. Krolak, J. Nelson (1978). *A Family of Truck Load Clustering Heuristics for Vechicle Routing Problems,* Technical report 78-2, Department of Computer Science, Vanderbilt University, Nashville, TN.

Gives an example for which both the savings and sweep algorithms for the VRP produce poor answers. A clustering based heuristic is then described which avoids some of these problems. The method is based on choosing the customer clusters by solving a capacitated facility location type problem, and then allocating vehicles to clusters. Good results are claimed on real problems but no comparisons are given with the other methods.

3.3. *Route first-cluster second*

J. Beasley (1983). Route first-cluster second methods for vehicle routing. *Omega 11,* 403-408.

Gives extensions to the basic route first-cluster second method of vehicle routing. Computational results are given for ten test problems from the literature.

R. Mole, D. Johnson, K. Wells (1983). Combinatorial analysis of route first-cluster second vehicle routing. *Omega 11,* 507-512.

Examines the reduction in the number of feasible solutions which follows the imposition of the constraint implied by the route first-cluster second methods.

M. Haimovich, A.H.G. Rinnooy Kan (to appear). Bounds and heuristics for capacitated routing problems: part I. *Math. Oper. Res.*

Considers the Euclidean VRP with vehicle capacities expressed in terms of the maximum number of customers that can be in any route. Develops upper and lower bounds on the optimal solution to this VRP which are asymptotically optimal. The heuristics used in the analysis correspond to partitioning a TSP tour through the customers. Other *area partitioning* heuristics are investigated for their worst case and probabilistic behavior and finally a polynomial *ϵ-approximation* heuristic is given which is not only of theoretical importance but may also be of practical significance.

3.4. *Improvement and exchange*

N. Christofides, S. Eilon (1969). An algorithm for the vehicle dispatching problem. *Oper. Res. Quart. 20,* 309-318.

Converts the VRP to a TSP with constraints and solves a thirteen-customer VRP exactly by branch and bound. Extends the 3-optimal method of Lin to the VRP, and gives computational results for this method on problems up to 100 customers. The test problems in this paper have become standard test problems in the VRP literature.

I. Cheshire, A. Malleson, P. Naccache (1982). A dual heuristic for vehicle scheduling. *J. Oper. Res. Soc. 33,* 51-61.

Initial routes satisfying 'local optimality' are modified to become feasible in a step-by-step procedure, where at each step the local optimality property is maintained.

3.5. *Mathematical programming*

(a) *(generalized) assignment*

M. Fisher, R. Jaikumar (1978). *A Decomposition Algorithm for Large-Scale Vehicle Routing*, Report 78-11-05, The Wharton School, University of Pennsylvania, Philadelphia.

Gives a formulation of the VRP in terms of two types of 0-1 variables: allocation variables of customers to routes, and routing variables of sequencing the customers within a route. A Benders decomposition procedure is suggested whereby the master problem involves the allocation stage (and can be solved as a generalized assignment problem with additional constraints), and the sub-problems involve the routing stage (and can be solved as TSPs). The method is capable of solving VRPs exactly, albeit of small size.

M. Fisher, R. Jaikumar (1981). A generalized assignment heuristic for vehicle routing. *Networks 11,* 109-124.

Describes a two-phase method which is, essentially, the first iteration of the Benders decomposition method by the same authors. In the first phase of assigning customers to routes, *m* customers are chosen as *seeds* to start the construction of the *m* route-clusters. Insertion costs are computed for all the other customers to indicate the cost of assigning a customer to a seed (and hence to the corresponding cluster). Using these costs a generalized assignment problem is solved to produce the route clusters. The second phase solves the TSP for each route cluster to produce the solution to the VRP. Very good computational results are reported for problems containing up to 200 customers.

B. Gavish, E. Shlifer (1979). An approach for solving a class of transportation scheduling problems. *European J. Oper. Res. 3,* 122-134.

An algorithm is developed for solving a wide class of problems in routing and scheduling, including the TSP, VRP, and school bus problem. These problems are formulated as assignment problems with additional constraints and are solved by a branch and bound procedure. Cases are mentioned when the assignment problem solutions are (or can be made) feasible - and optimal - without the need of the tree search. Computational results are given for some very large bus scheduling problems.

P. Van Leeuwen, A. Volgenant (1983). Solving symmetric vehicle routing problems asymmetrically. *European J. Oper. Res. 12,* 388-393.

The VRP with a symmetric travel matrix is transformed to one with an asymmetric matrix, on the principle that bounds from the assignment relaxation of TSP-type problems are better when the cost matrices are asymmetric. The problem is then relaxed, and the solution to this relaxed problem is 'adjusted' to produce a VRP solution. Some computational results are included.

(b) *set partitioning and set covering*

M.L. Balinski, R. Quandt (1964). On an integer program for a delivery problem. *Oper. Res. 12,* 300-304.

Gives a set covering formulation of the VRP, where variables correspond to the (enumerated) routes.

F. Cullen, J. Jarvis, D. Ratliff (1981). Set partitioning-based heuristics for interactive routing. *Networks 11,* 125-144.

A man-machine interactive approach is used for solving a class of routing problems including the VRP and the dial-a-ride problem. A set partitioning model forms the basis of the approach, together with a pricing mechanism for generating new routes (columns). The implementation on a color graphics terminal has produced good results on standard test problems.

B. Foster, D. Ryan (1976). An integer programming approach to the vehicle scheduling problem. *Oper. Res. Quart. 27,* 367-384.

Formulates the VRP as a set covering problem (where the columns represent routes) and a column generation procedure is suggested. The integer program is first solved as a linear program by the revised simplex method and information obtained from the linear program together with some route shape restrictions that are added, are used to generate new columns. Cuts are added to achieve integrality. Good computational results are given for fifteen test problems.

(c) *state space relaxation*

N. Christofides, A. Mingozzi, P. Toth (1981). State-space relaxation procedures for the computation of bounds to routing problems. *Networks 11,* 145-164.

Introduces *state space relaxation* (SSR), a procedure for reducing the state space of a dynamic programming recursion so that the relaxed recursion can be easily solved. Relaxation procedures are described so that the relaxed recursion provides a strict bound to the value of the original dynamic program. The procedure is analogous to (but more general than) Lagrangean relaxation and can provide bounds for branch and bound algorithms for combinatorial problems. SSR is explained for the TSP, time-constrained TSP and VRP. SSR is one of the few effective methods for the exact solution of the VRP.

N. Christofides, A. Mingozzi, P. Toth (1980). Exact algorithms for the vehicle routing problem based on spanning tree and shortest path relaxation. *Math. Programming 19,* 255-282.

Two methods are described for computing lower bounds to the VRP. The first is a simple Lagrangean bound based on the m-TSP. The second is a bound which can be derived from a Lagrangean relaxation of the set covering formulation of the VRP and the use of dominance considerations. This second bound and its generalizations are related to SSR and q-paths - the entities produced by the solution of the relaxed dynamic programming recursion. The q-path bound is inserted in a branch and bound and computational results are reported for the exact solution of VRPs up to 25 customers using this algorithm. (Since this paper was written, improvements to the algorithm has enabled VRPs with more than 50 customers to be solved.)

H. Trienekens (1982). *The Time Constrained Vehicle Routing Problem,* Unpublished report, Erasmus University, Rotterdam.

Considers the VRP with call time-window constraints at the customers, and for this problem derives lower bounds which are based on improvements and extensions of the q-route concept introduced in [Christofides, Mingozzi & Toth 1980] (see above). These bounds together with a heuristic upper bound are inserted in a tree search procedure which is then tested on eleven test problems from the literature (with the time constraints added).

3.6. *Miscellaneous*

S. Gorenstein (1970). Printing press scheduling for multi-edition periodicals. *Management Sci. B16,* 373-383.

Converts the scheduling problem on a printing press to an equivalent VRP, proposes a heuristic and includes an example.

W. Turner, E. Hougland (1975). The optimal routing of solid waste collection vehicles. *AIIE Trans. 7,* 427-431.

A modified m-TSP cost matrix is used to model a VRP representing a solid-waste collection system. A heuristic and two examples are discussed.

C.L. Doll (1980). Quick and dirty vehicle routing procedure. *Interfaces 10,* 84-85.

States that the current survey is unnecesary since no heuristic (let alone an optimization algorithm) can be of any use for the VRP!!

J. Bartholdi, L. Platzman, R. Collins, W. Warden (1983). A minimal technology routing system. *Interfaces 13,* 1-8.

Describes the implementation of a very simple route first-cluster second heuristic for the VRP, based on a heuristic for the TSP derived from space-filling curves. This system can operate manually and has achieved - in one case - a 13% saving compared to the previous manual system.

4. Variations of the VRP

4.1. *Multiperiod*

E. Beltrami, L. Bodin (1974). Networks and vehicle routing for municipal waste collection. *Networks 4,* 65-94.

Discusses various aspects of the problem of waste collection, including the period VRP, multidepot VRP, and ARP. A number of heuristics developed and applied by the authors for solving such problems are described.

R. Russell, W. Igo (1979). An assignment routing problem. *Networks 9,* 1-17.

Describes several heuristics for the period VRP, including generalizations of the savings algorithm and the m-tour algorithm for the VRP from [Russell 1977] (see §7). Computational results are given for three problems, one problem involving 775 calls.

O.M. Raft (1982). A modular algorithm for an extended vehicle scheduling problem. *European J. Oper. Res. 11,* 67-76.

Describes a five stage approach to multiperiod VRPs. The stages consist of: (i) Cluster customers into routes; (ii) Assign clusters to depots (if more than one depot exists); (iii) Assign vehicles to clusters; (iv) Allocate the vehicle-cluster combinations to the days of the period; (v) Route each cluster in detail. Each stage is solved separately and then the solutions are connected by an iterative procedure.

N. Christofides, J. Beasley (1984). Multiperiod routing problems, *Networks 14,* 237-256.

Introduces two *seed problems,* the p-median and the TSP, as 'relative order' approximations to the values of feasible solutions to the multiperiod VRP. The day combination of each customer is then chosen so that the value obtained by solving a seed problem (instead of the VRP) for each day of the period, is minimized. Once the calls are allocated to days, the VRP is solved for each day. Computational results are given for test problems.

4.2. *Multidepot*

E. Beltrami, N. Bhagat, L. Bodin (1971). *A Randomized Routing Algorithm with Application to Barge Dispatching in New York City.* Report 71-15, State University of New York, Stony Brook.

Extends the savings algorithm to a routing problem with multiple depots and where customers may have different call frequency requirements. The problem is a multidepot version of the period VRP. The problem arose in routing barges at the Department of Sanitation in New York City, and the extension of the method was based on a randomization procedure for selecting links from the savings list.

B. Gillett, J. Johnson (1976). Multiterminal vehicle dispatch algorithm. *Omega 4,* 711-718.

Extends the sweep algorithm for the VRP to the case when more than one depot exists. Computational results are included.

4.3. *Fleet size*

F. Gheysens, B. Golden, A. Assad (1982). *A Relaxation Heuristic for the Fleet Size and Mix Vehicle Routing Problem,* Working paper MS/S 82-029, University of Maryland, College Park.

A heuristic, based on Lagrangean relaxation, is presented for choosing a vehicle fleet size and mix so that the combined cost of vehicle acquisition and routing is minimized. Computational results are given for problems up to 75 customers.

B. Golden, A. Assad, L. Levy, F. Gheysens (1982). *The Fleet Size and Mix Vehicle Routing Problem,* Working paper MS/S 82-020, University of Maryland, College Park.

Describes several heuristics for choosing a vehicle fleet size and mix so that the combined cost of vehicle acquisition and routing is minimized. Extensive computational results are given. The results of the heuristics are compared with lower bounds and estimates of the optimal solutions.

A. Marchetti Spaccamela, A.H.G. Rinnooy Kan, L. Stougie (to appear). Hierarchical vehicle routing problems. *Networks.*

Considers the problem of acquiring a vehicle fleet so as to minimize the sum of fixed vehicle costs and variable running costs. It is assumed that the exact location of customers is unknown when the vehicle buying decision is made. It is also assumed that the variable cost is proportional to the length of the longest route (i.e., an even workload is preferred). A two-stage stochastic programming formulation is given, and it is shown that a simple heuristic has an asymptotically optimal behavior.

4.4. *Fixed routes*

N. Christofides (1971). Fixed routes and areas for delivery operations. *Internat. J. Physical Distribution,* 87-92.

A set of customer areas and the demand within each area are given for each day of a period. A heuristic is described which produces a set of 'fixed' routes passing through each customer area. These routes are required to be feasible for each of the days in the period. Once in an area, a vehicle is assumed to visit all the customers (and supply all the demand) within the area. Note that this problem is similar to the VRP with stochastic demands. An example is given.

J. Beasley (1984). Fixed routes. *J. Oper. Res. Soc. 35,* 49-55.

Considers the fixed routes problem, the problem of designing routes for delivery vehicles that can be operated unchanged for a given period of time. It shows how standard vehicle routing algorithms can be adapted to deal with the daily version of the problem. Computational results are presented for these adapted algorithms for a number of test problems from the literature.

4.5. *Dynamic demand*

W. Stewart, B. Golden, F. Gheysens (1982). *A Survey of Stochastic Vehicle Routing,* Working paper MS/S-82-027, University of Maryland, College Park.

Gives formulations and heuristic algorithms for a VRP with stochastic demand. The problem is to design routes so that the sum of the *route-length costs* and the *penalty* paid if any route becomes infeasible because of excessive demand is minimized. Note that this problem is similar to that of designing fixed routes.

4.6. *Inventory VRP*

I. Or (1983). *A Heuristic Solution Procedure for the Inventory Routing Problem,* Working paper MS/S 83-029, University of Maryland, College Park.

Some formulations of the inventory VRP are given and heuristic solution procedures suggested. No computational testing is reported.

A. Federgruen, P. Zipkin (to appear). A combined vehicle routing and inventory allocation problem. *Oper. Res.*

Gives a formulation of the inventory VRP and describes a heuristic solution procedure which is suggested by the Benders decomposition approach to the problem.

5. Arc Routing Problems

N. Christofides (1973). The optimum traversal of a graph. *Omega 1,* 719-732.

Gives a heuristic for the ARP and tests the heuristic on graphs up to 50 vertices. A lower bound is suggested (for comparison purposes) which is wrong, and was later corrected in [Golden & Wong 1981] (see below).

C. Orloff (1974A). A fundamental problem in vehicle routing. *Networks 4,* 35-64.

Gives a review of routing problems, including the TSP, CPP and RPP. Introduces a transformation of vertices to arcs (in an attempt to reduce the size of the graph to be traversed), which was later shown in [Lenstra & Rinnooy Kan 1976] (see below) to be wrong. Many of the ideas in this paper are, however, useful.

C. Orloff (1974B). Routing a fleet of *m* vehicles to/from a central facility. *Networks 4,* 147-162.

Gives a transformation of the *m*-vehicle VRP (or other routing problems) to a single-vehicle VRP (or other single-vehicle routing problems). The transformation was corrected in [Lenstra & Rinnooy Kan 1976] (see below).

J.K. Lenstra, A.H.G. Rinnooy Kan (1976). On general routing problems. *Networks 6,* 273-280.

Considers the problem of minimum cost traversal of a subset of vertices and arcs of a graph and shows that the conversion of required vertices to required arcs proposed in [Orloff 1974A] (see above) is incorrect.

H. Stern, M. Dror (1979). Routing electric meter readers. *Computers and Oper. Res. 6,* 209-223.

Gives a heuristic for solving the ARP and demonstrates the effectiveness of this heuristic by an application to a real-life problem.

M. Ball, L. Bodin, B. Golden, A. Assad, C. Stathes (1981). *A Strategic Truck Fleet Sizing Problem Analyzed by a Routing Heuristic,* Working paper MS/S-81-006, University of Maryland, College Park.

Gives three heuristic algorithms (two of the route first-cluster second variety, and the third of the 'minimum cost per unit delivered' variety), for solving a special case of the ARP. The demand associated with a directed arc in this ARP is specified in terms of integer truck loads required per week. The option of using a common carrier for certain shipments is considered. The application to a real-life problem is discussed.

B. Golden, R. Wong (1981). Capacitated arc routing problems. *Networks 11,* 305-315.

Gives mathematical formulations for the ARP and describes a heuristic for solving the problem. Gives computational results for a number of problems. Also corrects a simple lower bound suggested earlier in [Christofides 1973] (see above).

B. Golden, J. DeArmon, E. Baker (1982). *Computational Experiments with Algorithms for a Class of Routing Problems,* Working paper MS/S 81-033, University of Maryland, College Park.

Develops and tests three heuristic methods for the ARP. The computational results are extensive and represent the most significant investigation of the ARP to date.

6. Dial-a-Ride, School Bus Routing and Scheduling

P. Krolak, M. Williams (1978). *Computerized School Bus Routing,* Technical report 78-1, Department of Computer Science, Vanderbilt University,

Nashville, TN.

Describes a heuristic which solves the school bus routing problem by: (i) clustering the bus stops, (ii) assigning vehicles to clusters, (iii) optimal sequencing of stops within the clusters. Parts (i) and (ii) are achieved by solving transportation/assignment problems whereas part (iii) involves the heuristic solution of TSPs.

B. Gavish, K. Srikanth (1979). *Mathematical Formulations for the Dial-a-Ride Problem,* Working Paper 7909, Graduate School of Management, University of Rochester, NY.

Various versions of the dial-a-ride problem are formulated including the single-vehicle and multivehicle versions. Relationships with other routing problems are discussed.

H. Psaraftis (1980). A dynamic programming solution to the single-vehicle many-to-many immediate request dial-a-ride problem. *Transportation Sci. 2,* 130-154.

Gives an exact dynamic programming procedure for the dial-a-ride problem.

M. Ball, L. Bodin, R. Dial (1983). A matching based heuristic for scheduling mass transit crews and vehicles. *Transportation Sci. 17,* 4-31.

Describes a procedure for the simultaneous scheduling of vehicles and crews in a mass transit transport system. The solution of matching problems in graphs is used as the central algorithmic tool. The effectiveness of the heuristic is demonstrated by computational tests on real-life data involving up to 1000 trips.

J. Desrosiers, F. Soumis, M. Desrochers (1983). *Routing with Time-Windows by Column Generation,* Research report G-83-15, GERAD, Ecole des Hautes Etudes Commerciales, Québec.

Formulates the school bus routing problem as an m-TSP with visit time restrictions. In this formulation, a 'city' is a trip to be performed and the constraints representing the visit-time restrictions also act as the subtour elimination constraints of the TSP in much the same way as the traditional Miller-Tucker-Zemlin constraints. The algorithm uses column generation on a set partitioning problem, where news colunms are generated by solving shortest path problems and where the set partitioning problem is solved by linear programming and branch and bound. Extensive computational results are given for problems up to 151 trips.

7. THE m-TSP

M. Bellmore, S. Hong (1974). Transformation of the multisalesmen problem to the standard traveling salesman problem. *J. Assoc. Comput. Mach. 21,* 500-

504.
Gives a transformation of the graph on which the m-TSP is to be solved, so that only the solution of the TSP is needed on the transformed graph.

B. Gavish (1976). A note on 'The formulation of the M-salesman traveling salesman problem'. *Management Sci. 22,* 704-705.
Gives a formulation of the m-TSP.

R. Russell (1977). An effective heuristic for the m-tour TSP with some side conditions. *Oper. Res. 25,* 517-524.
Gives a heuristic for the VRP which is a generalization of the Lin-Kernighan heuristic for the TSP. The initial starting solution (which is then improved by the heuristic) is one obtained by the savings or sweep algorithms. Computational results are given for thirteen test problems up to 163 customers.

H. Stern, M. Dror (1977). *On the Multisalesmen Problem and Extensions,* Working paper 2/77, University of the Negev, Beer-Sheva.
Gives a survey and a classification of the m-TSP and its extensions. Four main classes of m-TSPs are used for this classification depending on whether the number m of salesmen is given or free, and on whether the location of the 'base city' for the salesmen is given or is to be chosen. The case of more than one 'base city' is also mentioned.

G. Laporte, Y. Nobert (1980). A cutting-plane algorithm for the m-salesmen problem. *J. Oper. Res. Soc. 31,* 1017-1024.
An integer programming formulation of the m-TSP is solved by linear programming and Gomory cutting planes. Subtour elimination constraints are added as and when needed. m-TSPs with up to 100 cities are reported solved in this way.

11

Sequencing and Scheduling

J.K. Lenstra
Centre for Mathematics and Computer Science, Amsterdam

A.H.G. Rinnooy Kan
Erasmus University, Rotterdam

CONTENTS

The theory of scheduling is concerned with the *optimal allocation* of scarce *resources* to *activities* over time. Of obvious practical importance, it has been the subject of extensive research over the past decades. In view of the fact that the above description allows for a wide variety of problem types, it should come as no surprise that the development of the theory has gone hand in hand with the refinement of a detailed *problem classification*, for which the ultimate foundation was laid in the classic book *Theory of Scheduling* [Conway, Maxwell & Miller 1967] (see §1.1).

Partly under the influence of their work, the emphasis has been on the investigation of *machine scheduling problems*, in which the activities are represented by *jobs* and the resources by *machines*, each of which can process at most one job at a time. Typically, the number of feasible allocations or *schedules* will be finite, but very large. If all the relevant information on jobs, machines and optimality criterion is known in advance, the scheduling problem becomes an example of a *combinatorial optimization* problem, and indeed most of the techniques developed for such problems have at some point been applied to scheduling problems.

One of the techniques that has been especially successful is the *complexity classification* that results from the theory of $\mathcal{NP}$-completeness (see [Garey & Johnson 1979], §1.1). This theory allows for a formal interpretation of the empirical difference between *easy* and *difficult* combinatorial optimization problems, by equating the former group with the problems that are *well solvable* in the sense that their solution requires only time bounded by a polynomial function of problem size, and the latter group with the *$\mathcal{NP}$-hard* problems for which a polynomial algorithm is very unlikely to exist.

The application of $\mathcal{NP}$-completeness theory in conjunction with various algorithmic techniques has succeeded in settling the complexity status ('well-solvable' or '$\mathcal{NP}$-hard') of the large majority of the scheduling problems that occur in a detailed problem classification first published in [Graham, Lawler, Lenstra & Rinnooy Kan 1979] (see §1.2). We trust that we will be excused for adhering closely to their classification in this bibliography.

Thus, we first classify scheduling problems according to the type of *machine environment* in which they are situated. The simplest such environment is a *single machine,* on which job j has to spend an (integral) *processing time* p_j ($j = 1,...,n$). An obvious generalization is to assume that each job has to be executed on any one of m *parallel machines,* which may be *identical, uniform* (machine i processes its jobs at *speed* s_i) or *unrelated* (machine i is able to process job j at speed s_{ij}). Another generalization is to assume that each job may have to visit more than one machine: if each job requires processing on all m machines in arbitrary order, the system is called an *open shop*; if each job has to visit all machines in a fixed order which is the same for each job, we have a *flow shop*; if the orders are fixed but possibly different for each job, we have a *job shop.* As the bibliography will partly reveal, flow shops and job shops have been the traditional domain of operations researchers and industrial engineers, whereas the study of parallel machine systems has been strongly influenced by their applicability in computer science.

Within each of these subclasses, we may further classify problem types by specifying certain *job characteristics.* First of all, it is important to distinguish between the case that *preemption* (job splitting) is allowed at zero cost and the case that a job, once started on a machine, must be processed without interruption until its completion on that machine. Secondly, various types of *precedence constraints* may be defined on the job set, that must be respected in each feasible schedule. Another way to generalize the model is to assume that job j becomes available for processing at an (integral) *release date* r_j and has to be completed no later than an (integral) *deadline* d_j; in the basic model, all $r_j = 0$ and all $d_j = \infty$. Further, it is often fruitful to consider the special case of *unit processing times,* in which all $p_j = 1$.

The final component of the classification scheme is the *optimality criterion* adopted. This is usually defined in terms of cost functions f_j of the job *completion times* C_j in a particular schedule; f_j may also depend on a given (integral) *due date* d_j and *weight* $w_j (j = 1,...,n)$. We distinguish between *minmax* criteria, i.e., the minimization of *maximum cost* $\max_j \{f_j(C_j)\}$, and *minsum* criteria, i.e., the minimization of *total cost* $\Sigma_j f_j(C_j)$. Important minmax criteria are *maximum completion time* $\max_j \{C_j\}$ and *maximum lateness* $\max_j \{C_j - d_j\}$; important minsum criteria are *total completion time* $\Sigma_j C_j$, *total tardiness* $\Sigma_j \max\{0, C_j - d_j\}$, the *number of late jobs* Σ_j (*if* $C_j \leq d_j$ *then* 0 *else* 1), and the *weighted* versions of these in which the j-th term is multiplied by w_j $(j = 1,...,n)$.

It should be apparent that the number of problems in the above class is huge. Still, as we shall see below, many interesting problem types are not included and require special introduction in the bibliography.

In drawing up this bibliography, we have concentrated on publications that appeared in 1981 or later. For the literature prior to 1981, we refer to the books and surveys listed in §§1.1,2. We have exercised some judgement in determining which publications to include; if any reader feels we have overlooked an important contribution, we would be pleased to hear from him or her. We have not included papers on *parallel* scheduling algorithms, as those are dealt with in one of the other contributions to this volume.

1. Books and Surveys

In this section, we list some books and surveys that will serve as a general introduction to the area and to the less than recent literature, as well as some papers that bear on the problem classification.

1.1. *Books*

R.W. Conway, W.L. Maxwell, L.W. Miller (1967). *Theory of Scheduling,* Addison-Wesley, Reading, MA.

As the first serious book on scheduling theory, this text is now rather outdated but still remarkable for the way it mixes deterministic scheduling with queueing and simulation - a mix that has recently become fashionable again (see §11).

K.R. Baker (1974). *Introduction to Sequencing and Scheduling,* Wiley, New York.

Although this textbook largely ignores recent issues such as computational complexity and analysis of heuristics, it does provide a very readable introduction to the basic results in the area.

E.G. Coffman, Jr. (ed.) (1976). *Computer & Job/Shop Scheduling Theory,* Wiley, New York.

This edited collection of papers contains some careful reviews of the state of the art around 1975, with particularly nice contributions by R. Sethi on minimizing maximum completion time and by R.L. Graham on the worst case analysis of heuristics.

A.H.G. Rinnooy Kan (1976). *Machine Scheduling Problems: Classification, Complexity and Computations,* Nijhoff, The Hague.
J.K. Lenstra (1977). *Sequencing by Enumerative Methods,* Mathematical Centre Tract 69, Centre for Mathematics and Computer Science, Amsterdam.

These Ph.D. theses contain surveys of optimization algorithms and complexity results. For some single machine, flow shop and job shop problems, branch-and-bound algorithms are developed and evaluated.

M.R. Garey, D.S. Johnson (1979). *Computers and Intractability: a Guide to the Theory of NP-Completeness,* Freeman, San Francisco.

The first textbook on computational complexity offers a well-written introduction to the tools and techniques in this area, with an extremely useful survey of $\mathcal{NP}$-completeness results at the end.

S. French (1982). *Sequencing and Scheduling: an Introduction to the Mathematics of the Job-Shop,* Horwood, Chichester.

Aimed at the same audience as [Baker 1974] (see above), this text covers most of the classical scheduling theory, including computational complexity and analysis of heuristics but with less emphasis on parallel machine models.

M.A.H. Dempster, J.K. Lenstra, A.H.G. Rinnooy Kan (eds.) (1982). *Deterministic and Stochastic Scheduling,* Reidel, Dordrecht.

The proceedings of the NATO Advanced Study and Research Institute on Theoretical Approaches to Scheduling Problems, held in Durham, England in 1981, provide some up-to-date surveys. Particular attention is paid to the interfaces between deterministic and stochastic scheduling. Among the contributors are E.G. Coffman, Jr., M.L. Fisher, J.C. Gittins, E.L. Lawler, M.L. Pinedo, S. Ross, L.E. Schrage, K. Sevcik and G. Weiss.

1.2. *Surveys*

M.J. Gonzalez, Jr. (1977). Deterministic processor scheduling. *Comput. Surveys 9,* 173-204.

A less than complete selection from the scheduling results available in 1977, aimed at a computer science audience.

R.L. Graham, E.L. Lawler, J.K. Lenstra, A.H.G. Rinnooy Kan (1979). Optimization and approximation in deterministic sequencing and scheduling: a survey. *Ann. Discrete Math. 5,* 287-326.

Written on the occasion of the DO77 conference in Vancouver in 1977, this survey provides a comprehensive review of optimization and approximation algorithms, including complexity results and worst case performance bounds, based on the problem classification sketched above. More than 150 references to the literature are listed. Need we say more?

E.L. Lawler, J.K. Lenstra, A.H.G. Rinnooy Kan (1982). Recent developments in deterministic sequencing and scheduling: a survey. M.A.H. Dempster, J.K. Lenstra, A.H.G. Rinnooy Kan (eds.). *Deterministic and Stochastic Scheduling,* Reidel, Dordrecht, 35-73.

On the occasion of the summer school on deterministic and stochastic scheduling in Durham, England in 1981, the preceding survey was revised to contain information on results up to 1981.

E.L. Lawler, J.K. Lenstra (1982). Machine scheduling with precedence contraints. I. Rival (ed.). *Ordered Sets,* Reidel, Dordrecht, 655-675.

Presented at the Symposium on Ordered Sets in Banff in 1981, this survey provides an exposition of the basic results in precedence constrained scheduling, including a treatment of the influence of so-called series-parallel constraints.

J.K. Lenstra, A.H.G. Rinnooy Kan (1984). Two open problems in precedence constrained scheduling. *Ann. Discrete Math.*

A contribution to the sequel of the Banff meeting, held in Lyon in 1982, this small survey deals with two open questions concerning scheduling unit-time jobs subject to precedence constraints, as well as a few new $\mathcal{NP}$-hardness results.

J. Carlier, P. Chrétienne (1982). Un domaine très ouvert: les problèmes d'ordonnancement. *RAIRO Rech. Opér. 16,* 175-217.

Written in French, this survey is in a sense an update of [Coffman 1976] (see §1.1), with emphasis on contributions by the authors. It is not so much a complete treatment as an attempt to focus on some of the main problem types and solution techniques.

E.L. Lawler (1983). Recent results in the theory of machine scheduling. A. Bachem, M. Grötschel, B. Korte (eds.). *Mathematical Programming: the State of the Art - Bonn 1982,* Springer, Berlin, 202-234.

A tutorial at the 11th International Symposium on Mathematical Programming in Bonn in 1982, this paper emphasizes polynomial algorithms and reviews a number of open problems.

E.G. Coffman, Jr., M.R. Garey, D.S. Johnson (1983). *Approximation Algorithms for Bin-Packing - an Updated Survey,* Bell Laboratories, Murray Hill, NJ.

This survey provides an overview of the analysis of approximation algorithms for the minimization of maximum completion time on identical parallel machines and for the related *bin packing* problem of minimizing the number of machines subject to a given bound on the maximum completion time.

D.S. Johnson (1983). The NP-completeness column: an ongoing guide. *J. Algorithms 4,* 189-203.

The seventh in a series of updates on [Garey & Johnson 1979] (see §1.1), this column surveys two types of complexity issues around parallel machine models: first the parallelization of algorithms and secondly the design and scheduling of multiprocessor systems.

S.C. Graves (1981). A review of production scheduling. *Oper. Res. 29,* 646-675.

This well-written survey deals with a wide range of sequencing and lot-sizing problems. A review of the practice of production scheduling leads to various challenging research questions. More than 100 references are given.

1.3. *Classification*

The problem classification sketched above was introduced in [Conway, Maxwell & Miller 1967] (see §1.1) and refined in [Graham, Lawler, Lenstra & Rinnooy Kan 1979] (see §1.2). In addition, the following papers are relevant.

B.J. Lageweg, J.K. Lenstra, E.L. Lawler, A.H.G. Rinnooy Kan (1982). Computer-aided complexity classification of combinatorial problems. *Comm. ACM 25,* 817-822.

A computer program is described that maintains a record of the known complexity results for a structured class of combinatorial problems. Given listings of well-solvable and $\mathcal{NP}$-hard problems, the program employs a reducibility relation defined on the class to classify each problem as easy, hard or open and to produce listings of the hardest easy problems, the easiest open ones, the hardest open ones and the easiest hard ones. The application of the program to a class of 120 single machine problems is demonstrated.

B.J. Lageweg, E.L. Lawler, J.K. Lenstra, A.H.G. Rinnooy Kan (1981). *Computer Aided Complexity Classification of Deterministic Scheduling Problems,* Report BW 138, Centre for Mathematics and Computer Science, Amsterdam.

This documents the results obtained by application of the above-mentioned program to a class of 4536 scheduling problems.

N. Hefetz, I. Adiri (1982). A note on the influence of missing operations on scheduling problems. *Naval Res. Logist. Quart. 29,* 535-539.

If the processing time of an operation is equal to zero, this can be interpreted to mean that the processing time is infinitesimally small, but also that the operation does not exist. These interpretations are by no means equivalent, as is demonstrated by various examples.

2. Single Machine Scheduling: Minmax Criteria

2.1. *Maximum lateness*

J. Carlier (1982). The one-machine sequencing problem. *European J. Oper. Res. 11,* 42-47.

Although the problem of minimizing maximum lateness on a single machine subject to release dates is $\mathcal{NP}$-hard, it possesses sufficient structure to make it reasonably well solvable in practical terms. A very efficient branch-and-bound algorithm was developed by McMahon & Florian (*Management Sci. 17* (1975), 782-792) and refined by Lageweg, Lenstra & Rinnooy Kan (*Statist. Neerlandica 30* (1976), 25-41). The author improves over this method by proposing a different branching rule.

J. Erschler, G. Fontan, C. Merce, F. Roubellat (1982). Applying new dominance concepts to job schedule optimization. *European J. Oper. Res. 11,* 60-66.
J. Erschler, G. Fontan, C. Merce, F. Roubellat (1983). A new dominance concept in scheduling n jobs on a single machine with ready times and due dates. *Oper. Res. 31,* 114-127.

Dominance results among schedules may be used in the obvious way to speed up enumerative procedures. These two papers introduce dominance based on the (r_j, d_j)-intervals, assuming that the objective is simply to meet all due dates. The jobs for which the (r_j, d_j)-interval is minimal under the partial order defined by inclusion, turn out to play an important role: they may be assumed to appear in order of nondecreasing r_j, and the jobs dominated by them in the partial order are, roughly speaking, spread around them.

M.R. Garey, D.S. Johnson, B.B. Simons, R.E. Tarjan (1981). Scheduling unit-time tasks with arbitrary release times and deadlines. *SIAM J. Comput. 10,* 256-269.

The special case in which all processing times are equal (or, equivalently, all $p_j = 1$ and the r_j and d_j need not be integral) has been open for a long time. In this situation, feasibility of the r_j and d_j can be tested in $O(n \log n)$ time by what amounts to repeated application of a dynamic version of Jackson's rule, which gives priority to the available jobs with the smallest d_j.

G.N. Frederickson (1983). Scheduling unit-time tasks with integer release times and deadlines. *Inform. Process. Lett. 16,* 171-173.

If, in the above problem, all $p_j = 1$ and the r_j and d_j are integral, then Jackson's rule solves the problem in $O(n \log n)$ time. Here, it is shown how an optimal schedule can actually be constructed in $O(n)$ time.

2.2. *Maximum cost*

K.R. Baker, E.L. Lawler, J.K. Lenstra, A.H.G. Rinnooy Kan (1983). Preemptive scheduling of a single machine to minimize maximum cost subject to release dates and precedence constraints. *Oper. Res. 31,* 381-386.

The problem described in the title is solved in $O(n^2)$ time by generalizing Lawler's algorithm for the case of equal release dates.

C.L. Monma (1980). Scheduling to minimize the maximum job cost. *Oper. Res. 28,* 942-951.

Let c_j indicate the amount of resource consumed (or, if $c_j < 0$, contributed) by job j. The problem is to find a job permutation π minimizing the maximum cumulative cost $\max_j \{f_{\pi(j)}(\sum_{i=1}^{j-1} c_{\pi(i)})\}$. This problem is shown to generalize various scheduling problems. An $\mathcal{NP}$-hardness proof and polynomial algorithms for special cases are presented.

C.L. Monma (1981). Sequencing with general precedence constraints. *Discrete Appl. Math. 3,* 137-150.
J.B. Sidney (1981). A decomposition algorithm for sequencing with general precedence constraints. *Math. Oper. Res. 6,* 190-204.

These papers study under what conditions certain job interchange techniques can cope with general precedence constraints. This typically results in polynomial solvability for series-parallel constraints and in less complete characterizations of optimality for more complicated structures.

3. Single Machine Scheduling: Minsum Criteria

3.1. *Optimization algorithms: dynamic programming*

In the dynamic programming approach to minimizing total cost on a single machine subject to precedence constraints, the minimum cost of scheduling a set of jobs is related to the minimum cost of all its subsets that are feasible with respect to the precedence constraints. The implementation by Baker & Schrage (*Oper. Res. 26* (1978), 111-120, 444-449), in which each feasible subset receives an integer label within a certain range, produced impressive computional results. The labeling is, however, not compact in the sense that, conversely, not every integer in the range corresponds to a feasible subset. Thus, there appeared to be room for further improvement.

E.L. Lawler (1979). *Efficient Implementation of Dynamic Programming Algorithms for Sequencing Problems*, Report BW 106, Centre for Mathematics and Computer Science, Amsterdam.

An alternative to the implementation scheme of Baker & Schrage is proposed. Time is proportional to n times the number of feasible sets generated, and space is proportional to n plus the maximum number of feasible sets of given size.

E.P.C. Kao, M. Queyranne (1982). On dynamic programming methods for assembly line balancing. *Oper. Res. 30,* 375-390.

Carefully designed experiments confirm that Lawler's scheme is computationally superior to the Baker-Schrage scheme.

R.N. Burns, G. Steiner (1981). Single machine scheduling with series-parallel precedence constraints. *Oper. Res. 29,* 1195-1207.

A compact labeling scheme is developed for the case that the precedence constraints are series-parallel.

G. Steiner (1984). Single machine scheduling with precedence constraints of dimension 2. *Math. Oper. Res. 9,* 248-259.

The compact labeling scheme from the previous paper is generalized to the case that the precedence constraints have dimension 2.

E.L. Lawler (1982). *Scheduling a Single Machine to Minimize the Number of Late Jobs*, Preprint, Computer Science Division, University of California, Berkeley.

Three results are presented. One is an $O(n \log n)$ algorithm that improves on an $O(n^2)$ method of Kise, Ibaraki & Mine (*Oper. Res. 26* (1978), 121-126). Another is an $O(n^6)$ dynamic programming algorithm for finding an optimal preemptive schedule subject to arbitrary release dates. Finally, the problem (with equal release dates) is shown to be $\mathcal{NP}$-hard when there are deadlines in addition to due dates.

3.2. *Optimization algorithms: branch-and-bound*

C.N. Potts, L.N. Van Wassenhove (1982). A decomposition algorithm for the single machine total tardiness problem. *Oper. Res. Lett. 1,* 177-181.

The problem of minimizing total tardiness on a single machine can be solved in $O(n^4 \Sigma p_j)$ (i.e., pseudopolynomial) time by a dynamic programming algorithm due to Lawler (*Ann. Discrete Math. 1* (1977), 331-342), which decomposes each problem into subproblems. The authors use a similar decomposition approach, but apply the Baker-Schrage method as soon as the subproblems get sufficiently small. Supported by additional dominance rules, the algorithm solves problems of up to 100 jobs.

C.N. Potts, L.N. Van Wassenhove (1983). An algorithm for single machine sequencing with deadlines to minimize total weighted completion time. *European J. Oper. Res. 12,* 379-387.

Lagrangean relaxation of the constraints $C_j \leqslant d_j$ is applied. The multipliers are constrained so that a simple heuristic for the original problem provides an optimal solution to the relaxed one.

A.M.A. Hariri, C.N. Potts (1983). An algorithm for single machine sequencing with release dates to minimize total weighted completion time. *Discrete Appl. Math. 5,* 99-109.

In the same spirit as the previous paper, the constraints $C_j \geqslant r_j + p_j$ are dualized. A dynamic version of Smith's rule (order the jobs in order of nonincreasing w_j / p_j) solves the relaxed problem.

L. Bianco, S. Ricciardelli (1982). Scheduling of a single machine to minimize total weighted completion time subject to release dates. *Naval Res. Logist. Quart. 29,* 151-167.

The same problem, a simpler lower bound, and more elaborate dominance conditions.

3.3. *Approximation algorithms*

Theoretically, the best possible heuristics are *fully polynomial approximation schemes,* which produce an ϵ-optimal schedule in time polynomial in problem size and $1/\epsilon$.

E.L. Lawler (1982). A fully polynomial approximation scheme for the total tardiness problem. *Oper. Res. Lett. 1,* 207-208.

This scheme applies the author's pseudopolynomial algorithm (*Ann. Discrete Math. 1* (1977), 331-342) to a problem with rescaled processing times. The running time is $O(n^7/\epsilon)$.

G.V. Gens, E.V. Levner (1981). Fast approximation algorithms for job sequencing with deadlines. *Discrete Appl. Math. 3,* 313-318.

This fully polynomial approximation scheme for the problem of minimizing the weighted number of late jobs requires $O(n^2 \log n + n^2/\epsilon)$ time. The emphasis is on the derivation of a tight lower bound so that ideas for the related knapsack problem can be fruitfully employed.

3.4. *Related models: due date selection*

Rather than taking due dates as given, the two papers below treat them as decision variables.

K.R. Baker, J.W.M. Bertrand (1981). A comparison of due-date selection rules. *AIIE Trans. 13,* 123-131.

The problem of minimizing the average due date $\Sigma d_j/n$ is investigated under the assumption that no job may be late and that d_j only depends on job parameters such as r_j and p_j that are known in advance.

S.S. Panwalkar, M.L. Smith, A. Seidmann (1982). Common due date assignment to minimize total penalty for the one machine scheduling problem. *Oper. Res. 30,* 391-399.

A polynomial algorithm is given for the minimization of a weighted sum of a common due date d, total tardiness, and total earliness $\Sigma_j \max\{0, d - C_j\}$.

3.5. *Related models: minimization of variance*

J.J. Kanet (1981). Minimizing variation of flow time in single machine systems. *Management Sci. 27,* 1453-1459.

This paper presents a simple algorithm for minimizing the total absolute difference of job completion times on a single machine, and a heuristic for the more difficult problem of minimizing variance of completion times.

J.J. Kanet (1981). Minimizing the average deviation of job completion times about a common due date. *Naval Res. Logist. Quart. 28,* 643-651.

The total absolute difference between job completion times and a common due date on a single machine can be minimized by a minor modification of the first method from the previous paper.

3.6. *Related models: minimization of the number of setups*

Given a schedule of precedence constrained jobs, a *setup* is said to occur when a job does not directly follow one of its immediate predecessors. To the large literature on this problem, the following papers have been added.

W.R. Pulleyblank (1984). On minimizing setups in precedence constrained scheduling. *Discrete Appl. Math.*

$\mathcal{NP}$-Hardness for the case of a bipartite precedence graph is established, and a polynomial algorithm for (again!) series-parallel constraints is given.

M.M. Syslo (1984). Minimizing the jump number for partially ordered sets: a graph-theoretic approach. *Order 1,* 7-19.

The results for the case of series-parallel constraints are derived in a different manner.

D. Duffus, I. Rival, P. Winkler (1982). Minimizing setups for cycle-free ordered sets. *Proc. Amer. Math. Soc. 85,* 509-513.

The obvious lower bound on the minimum number of setups, given by the Dilworth width minus 1, is shown to be tight for the case that the precedence graph contains no alternating cycles.

G. Gierz, W. Poguntke (1983). Minimizing setups for ordered sets: a linear algebraic approach. *SIAM J. Algebraic Discrete Meth. 4,* 132-144.

A lower bound that dominates the previous one is presented and shown to be tight for a class slightly more general than series-parallel constraints.

3.7. *Related models: two criteria*

Given two optimality criteria, the following papers deal with the determination of the set of Pareto-optimal points.

L.N. Van Wassenhove, L.F. Gelders (1980). Solving a bicriterion scheduling problem. *European J. Oper. Res. 4,* 42-48.

A pseudopolynomial algorithm is given for the total completion time and maximum lateness criteria.

L.N. Van Wassenhove, K.R. Baker (1982). A bicriterion approach to time/cost trade-offs in sequencing. *European J. Oper. Res. 11,* 48-54.

A procedure is developed for the maximum completion cost and total crashing cost criteria; the crashing cost of job j is given by $c_j(b_j - p_j)$, where p_j $(a_j \leqslant p_j \leqslant b_j)$ is a decision variable and a_j, b_j and c_j are known. The procedure is polynomial under additional assumptions on the completion cost functions.

K.S. Lin (1983). Hybrid algorithm for sequencing with bicriteria. *J. Optimization Theory Appl. 39,* 105-124.

A dynamic programming approach is presented for the total completion time and total tardiness criteria.

4. Nonpreemptive Parallel Machine Scheduling: Independent Jobs

4.1. *Optimization algorithms*

J.Y.-T. Leung (1982). On scheduling independent tasks with restricted execution times. *Oper. Res. 30,* 163-171.

The problem of minimizing maximum completion time on m identical parallel machines can be solved by dynamic programming in $O(\log p_{\max} \cdot \log m \cdot n^{2(k-1)})$ time, if the p_j can take on at most k different values.

B. Simons (1983). Multiprocessor scheduling of unit-time jobs with arbitrary release times and deadlines. *SIAM J. Comput. 12,* 294-299.

The m-machine generalization of the single machine problem solved in [Garey, Johnson, Simons & Tarjan 1981] (see §2.1) is considered. An algorithm with running time $O(n^3 \log\log n)$ is developed.

B. Simons (1982). On scheduling with release times and deadlines. M.A.H. Dempster, J.K. Lenstra, A.H.G. Rinnooy Kan (eds.). *Deterministic and Stochastic Scheduling,* Reidel, Dordrecht, 75-88.

This paper surveys polynomial algorithms and $\mathcal{NP}$-completeness results for scheduling equal-length jobs on one or more identical parallel machines subject to release dates and deadlines.

I. Meilijson, A. Tamir (1984). Minimizing flow time on parallel identical processors with variable unit processing time. *Oper. Res. 32,* 440-446.

A classical result states that the problem of minimizing total completion time on identical parallel machines can be solved by the SPT rule, assigning the jobs to machines in order of nondecreasing p_j. If the machines have a speed that increases over time, the SPT rule remains optimal; if the speed decreases, the problem becomes $\mathcal{NP}$-hard.

4.2. *Approximation algorithms: identical machines*

Unless stated otherwise, the papers in §§4.2−4 consider the minimization of maximum completion time.

In a *list scheduling* heuristic, the jobs are placed in a fixed list and, at each step, the earliest available machine is selected to process the first available job on the list.

J.O. Achugbue, F.Y. Chin (1981). Bounds on schedules for independent tasks with similar execution times. *J. Assoc. Comput. Mach. 28,* 81-99.

For arbitrary list scheduling, tight worst case relative error bounds as a function of $\rho = p_{\max}/p_{\min}$ are obtained. E.g., if $\rho \leqslant 3$, then the bound is equal to $2 - 1/3\lfloor m/3 \rfloor$ if $m \geqslant 6$, $17/10$ if $m = 5$ and $5/3$ if $m = 3, 4$.

B.L. Deuermeyer, D.K. Friesen, M.A. Langston (1982). Scheduling to maximize the minimum processor finish time in a multiprocessor system. *SIAM J. Algebraic Discrete Meth. 3,* 190-196.

For the unusual criterion of maximizing the minimum machine completion time, the LPT list scheduling heuristic, in which the jobs are listed in order of nonincreasing p_j, is shown to have a worst case ratio of 4/3. While the result is similar to Graham's result for minimizing maximum completion time (see, e.g., [Coffman 1976] in §1.1), the proof technique is quite different.

4.3. *Approximation algorithms: uniform machines*

Y. Cho, S. Sahni (1980). Bounds for list schedules of uniform processors. *SIAM J. Comput. 9,* 91-103.

It is known that both arbitrary list scheduling on identical machines and LPT list scheduling on uniform machines have a worst case ratio tending to 2 if m goes to infinity. Here, it is shown that for arbitrary list scheduling on uniform machines, the ratio is not bounded by a constant but increases not faster than $O(\sqrt{m})$.

D.K. Friesen, M.A. Langston (1983). Bounds for multifit scheduling on uniform processors. *SIAM J. Comput. 12,* 60-70.

The multifit heuristic, which involves repeated application of the first-fit-decreasing heuristic for bin packing to the packing of jobs in m intervals $[0,C_{max}]$, is extended to uniform machines and shown to have a worst case ratio between 1.341 and 1.4. This is the best ratio found so far for this model.

4.4. *Approximation algorithms: unrelated machines*

E. Davis, J.M. Jaffe (1981). Algorithms for scheduling tasks on unrelated processors. *J. Assoc. Comput. Mach. 28,* 721-736.

An adaptation of list scheduling is considered that incorporates a search for a relatively fast machine for each job. The worst case ratio is shown to be $O(\sqrt{m})$.

5. PREEMPTIVE PARALLEL MACHINE SCHEDULING: INDEPENDENT JOBS

5.1. *Optimization algorithms*

G. Schmidt (1983). *Preemptive Scheduling on Identical Processors with Time Dependent Availabilities*, Bericht 83-4, Fachbereich 20 Informatik, Technische Universität Berlin.

In case the machines are available only in certain given time intervals, the existence of a feasible preemptive schedule can be tested in polynomial time.

C. Martel (1982). Preemptive scheduling with release times, deadlines and due times. *J. Assoc. Comput. Mach. 29,* 812-829.

Polymatroidal network flow techniques are used to construct a preemptive schedule on uniform machines respecting release dates and meeting deadlines (if it exists) in $O(m^2n^4+n^5)$ time. The algorithm is combined with search techniques to minimize maximum lateness in polynomial time as well.

6. Parallel Machine Scheduling: Precedence Constrained Jobs

6.1. *Optimization algorithms: unit-time jobs*

The fundamental algorithmic results for scheduling precedence constrained unit-time jobs on m identical parallel machines so as to minimize maximum completion time are Hu's algorithm (*Oper. Res. 9* (1961), 841-848) for the case of tree-type constraints and various polynomial algorithms for the case of two machines. The complexity of the problem is open for every fixed number of machines greater than two. There are persistent rumors that these problems have turned out to be well solvable.

O. Marcotte, L.E. Trotter, Jr. (1984). An application of matroid polyhedral theory to unit-execution time, tree-precedence constrained job scheduling. W.R. Pulleyblank (ed.). *Progress in Combinatorial Optimization,* Academic Press, New York, 263-271.

Hu's algorithm is rederived from a minmax result due to Edmonds on covering the elements of a matroid (here, a transversal matroid on the jobs) by its bases (here, so-called feasible machine histories).

C.L. Monma (1982). Linear-time algorithms for scheduling on parallel processors. *Oper. Res. 30,* 116-124.

The generalization of Hu's algorithm to the problem of minimizing maximum lateness subject to intree constraints and some other scheduling problems are implemented to run in linear time by an adapted version of bucket sorting.

H.N. Gabow (1982). An almost-linear algorithm for two-processor scheduling. *J. Assoc. Comput. Mach. 29,* 766-780.

The two-machine problem with arbitrary precedence constraints is solved by an adaptation of Hu's algorithm in almost linear time ...

H.N. Gabow, R.E. Tarjan (1983). A linear-time algorithm for a special case of disjoint set union. *Proc. 15th Annual ACM Symp. Theory of Computing,* 246-251.

... and in strictly linear time.

K. Nakajima, J.Y.-T. Leung, S.L. Hakimi (1981). Optimal two processor scheduling of tree precedence constrained tasks with two execution times. *Performance Evaluation 1,* 320-330.

The two-machine problem with tree-type constraints and processing times equal to 1 or 2 is solved by a complicated $O(n \log n)$ algorithm. (For practical purposes, a heuristic due to Kaufman (*IEEE Trans. Comput. 23* (1974), 1169-1174) which has a worst case absolute error of 1, may be more attractive.)

M.R. Garey, D.S. Johnson, R.E. Tarjan, M. Yannakakis (1983). Scheduling opposing forests. *SIAM J. Algebraic Discrete Meth. 4,* 72-93.

The m-machine problem in which the precedence graph is the disjoint union of an inforest and an outforest is considered. If m is arbitrary, the problem is $\mathcal{NP}$-hard; if m is fixed, it is solvable in polynomial time; if $m = 2$, there is a linear time algorithm.

D. Dolev, M.K. Warmuth (1982). *Profile Scheduling of Opposing Forests and Level Orders*, Research report RJ 3553, IBM, San Jose, CA.

Opposing forests can be scheduled in $O(n^{2m-2}\log n)$ time; this improves over the above algorithm. Level orders, in which any two incomparable jobs with a common predecessor or successor have identical sets of predecessors and successors, can be scheduled in $O(n^{m-1})$ time; the case of arbitrary m is $\mathcal{NP}$-hard.

D. Dolev, M.K. Warmuth (1982). *Scheduling Flat Graphs*, Research report RJ 3398, IBM, San Jose, CA.

The theorems and background of the results in the above paper are presented.

D. Dolev, M.K. Warmuth (1984). Scheduling precedence graphs of bounded height. *J. Algorithms 5,* 48-59.

Precedence graphs in which the longest path has at most h arcs can be scheduled in $O(n^{h(m-1)+1})$ time. The case $h = 2$ is already $\mathcal{NP}$-hard.

E. Mayr (1981). *Well Structured Programs Are Not Easier to Schedule*, Report STAN-CS-81-880, Department of Computer Science, Stanford University.

The m-machine problem remains $\mathcal{NP}$-hard if the graph has a so-called hierarchical parallel structure.

6.2. *Optimization algorithms: preemptive scheduling*

E.L. Lawler (1982). Preemptive scheduling of precedence-constrained jobs on parallel machines. M.A.H. Dempster, J.K. Lenstra, A.H.G. Rinnooy Kan (eds.). *Deterministic and Stochastic Scheduling,* Reidel, Dordrecht, 101-123.

Some well-solvable problems involving the nonpreemptive scheduling of unit-time jobs turn out to have well-solvable counterparts involving the preemptive scheduling of jobs with arbitrary processing times. The latter problems include the minimization of maximum lateness on m identical machines subject to intree constraints and on two uniform machines subject to release dates and arbitrary precedence constraints. These results suggest a strong relationship between the two models.

6.3. *Approximation algorithms*

M. Kunde (1981). Nonpreemptive LP-scheduling on homogeneous multiprocessor systems. *SIAM J. Comput. 10,* 151-173.

In critical path list scheduling, the jobs are listed in order of nonincreasing total processing time of all jobs on the longest path starting at the job in question. This rule is investigated for the case of finding nonpreemptive schedules on identical machines subject to tree-type and chain-type precedence constraints. In the former case, the worst case ratio is $2-2/(m+1)$; in the latter case, the ratio tends to $5/3$ as m goes to infinity.

7. Parallel Machine Scheduling: Related Models

7.1. *Additional resource constraints*

The class of scheduling models known as *resource constrained project scheduling,* in which resources are of a more general nature than machines, has generated an impressive literature of its own. Virtually all these problems are $\mathcal{NP}$-hard in a very strong sense. Below, we list a few publications that appear to be on the borderline between the general class and the restricted class considered here.

The first four papers deal with unit-time jobs, arbitrary precedence constraints and the maximum completion time criterion.

E.L. Lloyd (1980). List scheduling bounds for UET systems with resources. *Inform. Process. Lett. 10,* 28-31.

There are m identical machines and l additional resources h of size R_h; job j requires r_{hj} units of resource h during its execution ($j = 1,...,n$; $h = 1,...,l$). Arbitrary list scheduling is shown to have a tight worst case ratio of $\min\{m, 2-1/m+\Sigma R_h(1-1/m)\}$.

E.L. Lloyd (1981). Coffman-Graham scheduling of UET task systems with 0-1 resources. *Inform. Process. Lett. 12,* 40-45.

Here, all $R_h = 1$ and all $r_{hj} \in \{0,1\}$. A generalization of the Coffman-Graham labeling algorithm (*Acta Inform. 1* (1972), 200-213) turns out to have a similar worst case behavior as arbitrary list scheduling.

E.L. Lloyd (1982). Critical path scheduling with resource and processor constraints. *J. Assoc. Comput. Mach. 29,* 781-811.

A complicated analysis shows that, for the model of [Lloyd 1980] (see above), the worst case ratio of a generalization of Hu's algorithm is bounded by a piecewise linear function of l and m.

J. Blazewicz, J.K. Lenstra, A.H.G. Rinnooy Kan (1983). Scheduling subject to resource constraints: classification and complexity. *Discrete Appl. Math. 5,*

11-24.

A detailed complexity classification is given for problems with identical or uniform machines and various types of resource constraints, parametrized according to l, R_h and $\max\{r_{hj}\}$, each of which is taken to be equal to 1 or to an arbitrary integral value.

In another common model, each machine i has its own resource (say, memory) of size R_i and can only process jobs whose resource requirements are no larger than R_i.

T.-H. Lai, S. Sahni (1981). *Preemptive Scheduling of a Multiprocessor System with Memories to Minimize* $L_{\max}$, Technical report 81-20, Computer Science Department, University of Minnesota, Minneapolis.

A network representation yields a preemptive schedule on identical machines minimizing maximum lateness in $O(n^3)$ time.

T.-H. Lai, S. Sahni (1982). *Preemptive Scheduling of Uniform Processors with Memory*, Technical report 82-5, Computer Science Department, University of Minnesota, Minneapolis.

Linear programming formulations are given for finding preemptive schedules on uniform machines minimizing maximum completion time and maximum lateness.

Two yet different models conclude this subsection.

E.L. Lloyd (1981). Concurrent task systems. *Oper. Res. 29,* 189-201.

Again unit-time jobs, arbitrary precedence constraints and the maximum completion time criterion. Job j requires q_j identical machines during its execution. The problem is well solvable for $m = 2$ and $\mathcal{NP}$-hard for $m \geqslant 3$. Arbitrary list scheduling has a worst case ratio $(2m - q_{\max})/(m - q_{\max} + 1)$.

J. Carlier, A.H.G. Rinnooy Kan (1982). Scheduling subject to nonrenewable-resource constraints. *Oper. Res. Lett. 1,* 52-55.

If the resources are actually consumed by the jobs (take, e.g., money) and the machine capacity is not binding ($m \geqslant n$), then minmax problems are well solvable, even if the amount of resource becomes available gradually over time.

7.2. *Periodic scheduling*

In (preemptive) periodic scheduling, each job j has a *period* ρ_j and is to be executed in each interval $(r_j + k\rho_j, d_j + k\rho_j)$ $(k = 0,1,2,...)$. On a single machine, the rule that gives priority to the available job with the closest deadline is known to construct a feasible schedule, if one exists.

J.Y.-T. Leung, M.L. Merrill (1980). A note on preemptive scheduling of periodic, real-time tasks. *Inform. Process. Lett. 11,* 115-118.

The problem of deciding feasibility is shown to be $\mathcal{NP}$-complete for each $m \geqslant 1$. The above priority rule for $m = 1$ turns out to provide an exponential method, in the sense that it is sufficient to verify whether feasibility has been achieved in a period equal to the least common multiple of the ρ_j; after which the schedule repeats itself.

E.L. Lawler, C.U. Martel (1981). Scheduling periodically occurring tasks on multiple processors. *Inform. Process. Lett. 12,* 9-12.

The last mentioned result is extended to the case of unrelated machines.

A.A. Bertossi, M.A. Bonuccelli (1983). Preemptive scheduling of periodic jobs in uniform multiprocessor systems. *Inform. Process. Lett. 16,* 3-6.

The Lawler-Martel algorithm above allows a more efficient implementation in the case of uniform machines.

J.Y.-T. Leung, J. Whitehead (1982). On the complexity of fixed-priority scheduling of periodic, real-time tasks. *Performance Evaluation 2,* 237-250.

The case of identical machines and equal release dates is solved in pseudo-polynomial time. The complexity of this problem is still open.

7.3. *Restricted starting times*

K. Nakajima, S.L. Hakimi (1982). Complexity results for scheduling tasks with discrete starting times. *J. Algorithms 3,* 344-361.

A detailed complexity analysis is given for the problem of finding a feasible nonpreemptive schedule on m identical machines in which each job j may start at any one of k_j given starting times. Even if the processing times can assume only two different values, the problem turns out to be $\mathcal{NP}$-complete in the case that $m = 1$ and all $k_j \leqslant 3$ and in the case that m is arbitrary and all $k_j = 2$. For some more restricted cases, polynomial algorithms are developed.

K. Nakajima, S.L. Hakimi, J.K. Lenstra (1982). Complexity results for scheduling tasks in fixed intervals on two types of machines. *SIAM J. Comput. 11,* 512-520.

The problem is to find a nonpreemptive schedule on two types of parallel machines: inexpensive slow machines and expensive fast ones. Job j requires a processing time p_j on a slow machine or $q_j < p_j$ on a fast one. Two models are considered: (a) each job j must be processed in an interval $(r_j, r_j + p_j]$; (b) each job j must start at time r_j. The objective is to minimize total machine cost. Both problems turn out to be $\mathcal{NP}$-hard. For some special cases, in which all $q_j = 1$, polynomial algorithms are presented.

8. Open Shop Scheduling

8.1. *Optimization algorithms*

E.L. Lawler, J.K. Lenstra, A.H.G. Rinnooy Kan (1981). Minimizing maximum lateness in a two-machine open shop. *Math. Oper. Res. 6,* 153-158.
E.L. Lawler, J.K. Lenstra, A.H.G. Rinnooy Kan (1982). Erratum. *Math. Oper. Res. 7,* 635.
The problem of finding a preemptive schedule minimizing maximum lateness in a two-machine open shop is solved by a linear time algorithm. The nonpreemptive case is shown to be $\mathcal{NP}$-hard.

Y. Cho, S. Sahni (1981). Preemptive scheduling of independent jobs with release and due times on open, flow and job shops. *Oper. Res. 29,* 511-522.
The existence of a preemptive schedule respecting release dates and deadlines in an m-machine open shop can be determined by linear programming. The analogous problems for two-machine flow and job shops are $\mathcal{NP}$-hard.

T. Fiala (1983). An algorithm for the open-shop problem. *Math. Oper. Res. 8,* 100-109.
In a very original contribution, results from graph theory are invoked to show that the problem of finding a nonpreemptive schedule minimizing maximum completion time in an m-machine open shop can be solved in $O(m^3n^2)$ time if $\max_i\{\Sigma_j p_{ij}\} \geqslant (16m'\log m' + 5m')p_{\max}$, where m' is the roundup of m to the closest power of 2.

E.L. Lawler, M.G. Luby, V.V. Vazirani (1982). Scheduling open shops with parallel machines. *Oper. Res. Lett. 1,* 161-164.
For a generalization of the preemptive open shop problem, in which there are given speeds s_{ijk} at which machine i can process the kth operation of job j, a linear programming formulation minimizes maximum completion time.

8.2. *$\mathcal{NP}$-hardness results*

J.O. Achugbue, F.Y. Chin (1982). Scheduling the open shop to minimize mean flow time. *SIAM J. Comput. 11,* 709-720.
The problem of finding a nonpreemptive schedule minimizing total completion time in a two-machine open shop, so far a prominent open problem, is shown to be $\mathcal{NP}$-hard through a reduction starting form 3-PARTITION. Further, tight bounds are derived on the quality of arbitrary schedules and shortest-processing-time-first schedules for an m-machine open shop.

T. Gonzalez (1982). Unit execution time shop problems. *Math. Oper. Res. 7,* 57-66.
The problem of finding a nonpreemptive or preemptive schedule

minimizing total completion time in an m-machine open shop is shown to be $\mathcal{NP}$-hard, even if $p_{ij} \in \{0,1\}$ for all (i,j). Similar results hold for the problems of minimizing maximum or total completion time in flow and job shops.

9. Flow Shop Scheduling

9.1. *Optimization algorithms and $\mathcal{NP}$-hardness results*

F.Y. Chin, L.-L. Tsai (1981). On J-maximal and J-minimal flow-shop schedules. *J. Assoc. Comput. Mach. 28,* 462-476.

For special cases of the problem of minimizing maximum completion time in an m-machine flow shop in which, for some machine h, $p_{hj} = \max_i \{p_{ij}\}$ for all j or $p_{hj} = \min_i \{p_{ij}\}$ for all j, $\mathcal{NP}$-hardness results complemented by polynomial algorithms are derived. In addition, bounds on the length of arbitrary permutation schedules are derived.

J.O. Achugbue, F.Y. Chin (1982). Complexity and solution of some three-stage flow shop scheduling problems. *Math. Oper. Res. 7,* 532-544.

A detailed analysis of the three-machine flow shop problem, in which each machine may be maximal or minimal in the above sense, leads to an exhaustive complexity classification.

W. Szwarc (1981). Precedence relations of the flow-shop problem. *Oper. Res. 29,* 400-411.

Conditions are provided under which Johnson's algorithm for the two-machine flow shop can be extended to the m-machine case.

W. Szwarc (1983). Flow shop problems with time lags. *Management Sci. 29,* 477-481.

An extension of the flow shop model is shown to cover many flow shop problems with time lags. Application of Johnson's algorithm yields lower and upper bounds.

J. Grabowski (1982). A new algorithm of solving the flow-shop problem. G. Feichtinger, P. Kall (eds.). *Operations Research in Progress,* Reidel, Dordrecht, 57-75.

A new branching scheme is proposed for the permutation flow shop problem based on an analysis of the transformations required to shorten the critical path corresponding to the feasible schedule at the current node of the search tree, and as such related to earlier work by Balas (*Oper. Res. 17* (1969), 941-957). The algorithm uses the bounding scheme developed by Lageweg, Lenstra & Rinnooy Kan (*Oper. Res. 26,* (1978), 53-67). Grabowski's method requires less time and generates smaller search trees than the method of Lageweg *et al.*

J. Grabowski, E. Skubalska, C. Smutnicki (1983). On flow shop scheduling with release and due dates to minimize maximum lateness. *J. Oper. Res. Soc. 34,* 615-620.

The above approach is extended to the minimization of maximum lateness subject to release dates.

9.2. *Approximation algorithms*

I. Bárány (1981). A vector-sum theorem and its application to improving flow shop guarantees. *Math. Oper. Res. 6,* 445-452.

A surprising geometrical argument leads to a flow shop heuristic that requires $O(m^3n^2+m^4n)$ time and whose absolute error is bounded by $(m-1)(3m-1)p_{max}/2$. A remarkable feature of this result is that the error does not depend on n.

H. Röck, G. Schmidt (1982). *Machine Aggregation Heuristics in Shop Scheduling,* Bericht 82-11, Fachbereich 20 Informatik, Technische Universität Berlin.

Aggregation heuristics proceed by replacing m machines by two machines, on which the job processing times are given by the appropriate sums of the original processing times. The worst case ratios of such heuristics are proportional to m.

C.N. Potts (1981). *Analysis of Heuristics for Two-Machine Flow-Shop Sequencing Subject to Release Dates,* Report BW 150, Centre for Mathematics and Computer Science, Amsterdam.

For the problem of minimizing maximum completion time in a two-machine flow shop subject to release dates, three heuristics with worst case ratio 2 are presented. Repeated application of one of them, that is inspired by a dynamic application of Johnson's algorithm to a modified version of the problem, reduces the worst case ratio to 5/3.

9.3. *Related models: no wait in process*

There has always been a special interest in the flow shop model in which all operations of a job must be performed without interruption. The problem of minimizing maximum completion time under this restriction is a special case of the traveling salesman problem. The case of two machines is solvable in $O(n \log n)$ time by the Gilmore-Gomory algorithm for a special TSP; the case of four machines was proved $\mathcal{NP}$-hard by Papadimitriou & Kanellakis (*J. Assoc. Comput. Mach. 27* (1980), 533-549).

H. Röck (1984). The three-machine no-wait flow shop is NP-complete. *J. Assoc. Comput. Mach. 31,* 336-345

This settles the open question implied by the paragraph above.

H. Röck (1984). Some new results in flow shop scheduling. *Z. Oper. Res. 28,*

1-16.

The problems of minimizing maximum lateness and total completion time in a two-machine no wait flow shop are shown to be $\mathcal{NP}$-hard. For the case of unit processing times and a single additional resource of unit size, an $O(n \log n)$ time algorithm is presented.

10. Job Shop Scheduling

10.1. *Optimization algorithms*

The problem of minimizing maximum completion time in a job shop is $\mathcal{NP}$-hard, even in the case of three machines and unit processing times and in the case of two machines and processing times equal to 1 or 2 (Lenstra & Rinnooy Kan, *Ann. Discrete Math. 4* (1979), 121-140). Below, N denotes the total number of operations of all jobs.

N. Hefetz, I. Adiri (1982). An efficient optimal algorithm for the two-machines unit-time jobshop schedule-length problem. *Math. Oper. Res. 7,* 354-360.

The above problem with two machines and unit processing times is shown to be solvable in $O(N)$ time, through a rule that gives priority to the longest remaining job.

P. Brucker (1981). Minimizing maximum lateness in a two-machine unit-time job shop. *Computing 27,* 367-370.

In the same model, maximum lateness can be minimized in $O(N \log N)$ time; the priority of a job now depends on the difference between its due date and its number of operations.

P. Brucker (1982). A linear time algorithm to minimize maximum lateness for the two-machine, unit-time, job-shop, scheduling problem. R.F. Drenick, F. Kozin (eds.). *System Modeling and Optimization,* Lecture Notes in Control and Information Sciences 38, Springer, Berlin, 566-571.

The previous algorithm can be implemented to run in $O(N)$ time.

M.L. Fisher, B.J. Lageweg, J.K. Lenstra, A.H.G. Rinnooy Kan (1983). Surrogate duality relaxation for job shop scheduling. *Discrete Appl. Math. 5,* 65-75.

As part of the continuing (and, so far, rather fruitless) attack on the general job shop problem, computational experience is reported with surrogate duality relaxations of capacity and precedence constraints. Although the lower bounds dominate the classical ones and also those obtained by Lagrangean relaxation, a lot of time is required for their computation. The notorious 10-job 10-machine problem remains unsolved.

11. Probabilistic Scheduling Models

Probability theory finds application in scheduling in two ways. The first one is through a *probabilistic analysis* of the performance of scheduling rules: given a probability distribution over the class of problem instances, the behavior of a random variable representing the performance is investigated. The second way arises when certain job data are no longer assumed to be known in advance; for example, the processing time of a job may be a random variable, whose realization becomes known at the job's completion. The term *stochastic scheduling* is usually reserved for the latter interpretation. We list a few typical references in both areas. Ch.6, §7.3, gives more references on the probabilistic analysis of scheduling algorithms.

11.1. *Probabilistic analysis*

P.G. Gazmuri (1981). *Probabilistic Analysis of a Machine Scheduling Problem,* Unpublished manuscript.

The problem of minimizing total completion time on a single machine subject to release dates is studied under the assumption that processing times as well as release dates are independent and identically distributed. For each of two cases characterized by the relation between expected processing time and expected interarrival time, a heuristic is developed whose relative error tends to 0 in probability.

E.G. Coffman, Jr., G.N. Frederickson, G.S. Lueker (1982). Probabilistic analysis of the LPT processor scheduling heuristic. M.A.H. Dempster, J.K. Lenstra, A.H.G. Rinnooy Kan (eds.). *Deterministic and Stochastic Scheduling,* Reidel, Dordrecht, 319-331.

The average performance of the longest-processing-time-first rule, used to minimize maximum completion time on m identical parallel machines, is studied under the assumption that processing times are uniformly distributed on (0,1]. The ratio of expected LPT schedule length to expected optimal length is bounded by $1+O(m^2/n^2)$.

11.2. *Stochastic scheduling*

G. Weiss (1982). Multiserver stochastic scheduling. M.A.H. Dempster, J.K. Lenstra, A.H.G. Rinnooy Kan (eds.). *Deterministic and Stochastic Scheduling,* Reidel, Dordrecht, 157-179.

This is a survey of stochastic scheduling results for parallel machine models. Typical examples are the optimality of the longest(shortest)-expected-processing-time rule for minimizing maximum (total) completion time on uniform machines, under a variety of assumptions on the distribution of processing times.

M. Pinedo, L. Schrage (1982). Stochastic shop scheduling: a survey. M.A.H. Dempster, J.K. Lenstra, A.H.G. Rinnooy Kan (eds.). *Deterministic and Stochastic Scheduling,* Reidel, Dordrecht, 181-196.

This survey deals with stochastic scheduling results for open shop, flow shop (including the no wait case) and job shop models. Most of the stronger results are for two-machine shops. Much work remains to be done.

12. Related Scheduling Models

This final section is devoted to two scheduling models that do not fit into the preceding framework.

12.1. *Cyclic scheduling*

The (k,m)-cyclic staff scheduling problem is to minimize the number of workers in an m-period cyclic schedule such that requirements varying over the periods are met and each person works for k consecutive periods. In the obvious integer programming formulation, the coefficient matrix has a special structure that is capitalized on in the following papers.

J.J. Bartholdi III, H.D. Ratliff (1978). Unnetworks, with applications to idle time scheduling. *Management Sci. 24,* 850-858.

The complement of the matrix has exactly $m-k$ ones in each column. On the basis of this observation, the (5,7)-problem and several related problems are solved in polynomial time by a series of network flow or matching problems.

J.J. Bartholdi III, J.B. Orlin, H.D. Ratliff (1980). Cyclic scheduling via integer programming with circular ones. *Oper. Res. 28,* 1074-1085.

The (k,m)-problem is solved by transforming the integer program to a series of network flow problems. An unusual round-off property allows the problem also to be solved as a linear program. These techniques are extended to more general cyclic scheduling problems.

J.J. Bartholdi III (1981). A guaranteed-accuracy round-off algorithm for cyclic scheduling and set covering. *Oper. Res. 29,* 501-510.

If the workers are only intermittently available, the cyclic staff scheduling problem turns out to be $\mathcal{NP}$-hard, but the linear-programming round-off technique has an acceptable worst case absolute error.

12.2. *Hierarchical scheduling*

Often, scheduling is the last step in a sequence of planning decisions, where each decision affects the form and the constraints of its successors. If resources have to be acquired under uncertainty about what will be required of them, multistage stochastic integer programming formulations in which the scheduling decision appears at the last stage are a natural class of models. In the following papers, heuristics with strong properties of asymptotic optimality are developed for such models. The probabilistic analyses in question are based on accurate estimates of the value of an optimal schedule.

M.A.H. Dempster, M.L. Fisher, L. Jansen, B.J. Lageweg, J.K. Lenstra, A.H.G. Rinnooy Kan (1981). Analytical evaluation of hierarchical planning systems. *Oper. Res. 29,* 707-716.

This introductory paper provides the motivation for the approach sketched above and gives some preliminary results.

M.A.H. Dempster, M.L. Fisher, L. Jansen, B.J. Lageweg, J.K. Lenstra, A.H.G. Rinnooy Kan (1983). Analysis of heuristics for stochastic programming: results for hierarchical scheduling problems. *Math. Oper. Res. 8,* 525-537.

Results are presented for the case that the scheduling problem involves the minimization of maximum completion time on a set of identical or uniform parallel machines that has to be acquired when only the number of jobs and the probability distribution of their processing times are known.

M.A.H. Dempster (1982). A stochastic approach to hierarchical planning and scheduling. M.A.H. Dempster, J.K. Lenstra, A.H.G. Rinnooy Kan (eds.). *Deterministic and Stochastic Scheduling,* Reidel, Dordrecht, 271-296.

This paper includes a survey of relevant results in stochastic scheduling and discusses some interesting open questions.

J.B.G. Frenk, A.H.G. Rinnooy Kan, L. Stougie (1984). A hierarchical scheduling problem with a well-solvable second stage. *Ann. Oper. Res. 1.*

Here, the scheduling problem involves the minimization of total completion time.

J.K. Lenstra, A.H.G. Rinnooy Kan, L. Stougie (1984). A framework for the probabilistic analysis of hierarchical planning systems. *Ann. Oper. Res. 1.*

Relations between various measures of asymptotic optimality are derived, and general conditions are established under which a two-stage heuristic is asymptotically clairvoyant with probability 1.

12

Software

S. Powell
London School of Economics

CONTENTS

There is a continuum of problems from pure combinatorial problems, e.g. the set partitioning problem, the quadratic assignment problem, and pure integer programming (PIP) problems to mixed integer programming (MIP) problems. Many MIP problems are essentially large linear programming problems with a few integer variables.

Generally speaking commercial organizations who are manufacturing general software have concentrated their efforts on the MIP end of the spectrum. Such software is marketed as part of their complete mathematical programming system. As these codes are costly to produce, update and maintain they are aimed at a wide a market as possible. With the exception of PIPEX (see §2) all the commercial codes are MIP codes which can be used for the full range of problems.

The academic world has devoted its theoretical skills principally to pure combinatorial optimization. Frequently this work has been tested by writing software but the purpose of such a code is to enable the author to evaluate the theoretical development, not to produce a generally available code.

Many commercial organizations, e.g. airlines, bus companies and distributions companies, have developed codes for specific combinatorial problems. The heart of the code is often an algorithm, optimizing or heuristic, for a well-known combinatorial problem, e.g. the traveling salesman problem. The code is tailored to a particular customer to take account of 'complicating' constraints. These side constraints which may or may not be expressible as linear constraints, tend to destroy the special structure.

The only generally available codes which are maintained are those produced by commercial organizations; so availability in this context usually means at a cost. These codes were surveyed in [Land & Powell 1979] (see §1). Some of the codes in that survey are still in use but are no longer being developed, those referred to in §2 are currently under development.

All these commercial codes are basically linear programming codes with a branch and bound facility 'added on' (see §2). They all solve MIP problems except PIPEX which only solves zero-one problems. The enormous amount of research effort that has gone into developing other methods for integer programming is barely represented in the commercial codes. However, there are differences between the codes which do contain pointers to the future.

Apex IV and PIPEX reformulate the problem before attempting to solve it. This reformulation aims by logical analysis to reduce the size of the problem to be solved. PIPEX generates extra constraints derived from the facets of the knapsack problem to further tighten the problem before using branch and bound to solve the problem.

All commercial codes have special ordered sets of type 1 (S1) and most have special ordered sets of type 2 (S2) to represent nonlinear separable functions. The handling of nonlinear functions has been further enhanced by linked ordered sets and chains of linked ordered sets (see §5.1).

All the codes are essentially 'black boxes' which may be controlled by setting parameters before running the program. The do-it-yourself version of Sciconic gives the user internal access so that subroutiness can be written which use the data structures.

This bibliography refers the reader to the literature that is the immediate source of the ideas that are used in the currently available codes.

The traveling salesman problem is of interest not only in its own right but because it is used in many practical scheduling situations. A lot of work by many researchers has been devoted to this combinatorial problem and recent work (see §6) shows that it is possible to solve large problems. However, despite the interest in the problem and a solution method, to my knowledge there is no available code for the traveling salesman problem.

1. SURVEY

A. Land, S. Powell (1979). Computer codes for problems of integer programming. *Ann. Discrete Math. 5*, 221-269.

Presented at the DO77 conference in Vancouver in 1977, this is a survey of all, commercial and academic, computer codes that were available in 1977. Many of those codes are still being used. The survey desribes in detail the algorithms used by and the 'bells and whistles' present in the commercial codes.

2. Computer Manuals

This section refers to the computer manuals of the codes that are under current development.

Apex IV Reference Manual 84002550 Level B, Control Data Corporation, Publications Divisions, HQ02C, P.O. Box 0, Minneapolis, MN 55440, U.S.A.

LAMPS User Guide, Cap Scientific Ltd., 233 High Holborn, London WCIV 7DJ, U.K.

IBM Mathematical Programming System Extended/370 (MPSX/370), Mixed Integer Programming/370 (MIP/370): Program Reference Manual, Program Numbers 5740-XM3, 5746-XM2, IBM Corporation, Technical Publications, Department 824, 1133 Westchester Avenue, White Plains, NY 10604, U.S.A.

Integer Programming (IP2900) System, International Computers Ltd., ICL House, Putney, London SW15 lSW, U.K.

Mathematical Programming System MPSIII, Ketron Ltd., MSS Division, 18th Floor, Rosslyn Center, 1700 Moore Street, Arlington, VA 22209, U.S.A.

Mathematical Programmming System User Guide, Honeywell Information Systems, 60 Walnut Street, Wellesley Hills, MA 021181, U.S.A.

Pure Integer Programming/Executor (PIPEX), Description/Operations, IFP 5785-GBX, 1982, IBM Corporation, Technical Publications, Department 824, 1133 Westchester Avenue, White Plains, NY 10604, U.S.A.

Sciconic User Guide, Scicon Computer Services Ltd., Brick Close, Kiln Farm, Milton Keynes MK11, 3EJ, U.K.

3. Background to Branch and Bound

A. Land, A. Doig (1960). An automatic method of solving discrete programming problems. *Econometrica 28,* 497-520.

This is the original paper on branch and bound.

N.J. Driebeek (1966). An algorithm for the solution of mixed integer programming problems. *Management Sci. 12,* 576-587.

The first computer implementation of a branch and bound algorithm. It introduced dichotomous branching which has been used since in every commercial code.

A.M. Geoffrion, R.E. Marsten (1972). Integer programming algorithms: a

framework and a state-of-the-art survey. *Management Sci. 18,* 465-491.
As the title suggests this paper provides a terminology to describe integer programming algorithms.

4. Implementation

This group of papers describes, with computational experience, the algorithmic features that have been implemented.

M. Benichou, J.-M. Gauthier, P. Girodet, G. Hentges, G. Ribière, O. Vincent (1971). Experiments in mixed integer linear programming. *Math. Programming 1,* 76-94.

J.-M. Gauthier, G. Ribière (1977). Experiments in mixed-integer programming using pseudo-costs. *Math. Programming 12,* 26-47.
This is a continuation of the work reported by Benichou *et al.*; it includes a description of an implementation of S1 sets.

J.J.H. Forrest, J.P.H. Hirst, J.A. Tomlin (1974). Practical solutions of large mixed integer programming problems with UMPIRE. *Management Sci. 20,* 736-773.

G. Mitra (1973). Investigation of some branch and bound strategies for the solution of mixed integer linear programs. *Math. Programming 4,* 155-170.

J.A. Tomlin (1970). Branch and bound methods for integer and non-convex programming. J. Abadie (ed.). *Integer and Nonlinear Programming,* North-Holland, Amsterdam.
As well describing a branch and bound implementation for integer variables there is a description of an implementation of S1 and S2 sets.

5. Further Features

5.1. *Special ordered sets*

E.M.L. Beale, J.A. Tomlin (1970). Special facilities in a general mathematical programming system using ordered sets of variables. J. Lawrence (ed.). *Proc. 5th Internat. Conf. Operations Research,* Tavistock Publications, 447-454.
S1 and S2 sets were first described here.

E.M.L. Beale, J.J.H. Forrest (1976). Global optimization using special ordered sets. *Math. Programming 10,* 52-69.
This paper describes the automatic generation of variables in an S2 set which is representing a nonlinear function.

E.M.L. Beale (1979). Branch and bound methods. *Ann. Discrete Math. 5,* 201-219.

Linked ordered sets to represent product terms in a function are described here.

E.M.L. Beale, R.C. Daniel (1979). *Chains of Linked Ordered Sets: a New Formulation for Product Terms,* presented at the 10th International Symposium on Mathematical Programming, 1977; available from the authors at Scicon Ltd., Brick Close, Kiln Farm, Milton Keynes MK11 3EJ, U.K.

5.2. *Reformulation and constraint generation*

L.A. Oley, R.J. Sjoquist (1983). Automatic reformulation of mixed and pure integer models to reduce solution time in Apex IV. *SIGMAP Bull. 32,* 39-51.

H. Crowder, E.L. Johnson, M.W. Padberg (1983). Solving large-scale zero-one linear programming problems. *Oper. Res. 31,* 803-834.

This paper describes the algorithm used in PIPEX (see § 2); it consists of problem reformulation, followed by constraint generation and then branch and bound. Computional experience with some medium to large problems is reported.

6. A FUTURE CODE?

H. Crowder, M.W. Padberg (1980). Solving large-scale symmetric travelling salesman problems to optimality. *Management Sci. 26,* 495-509.

An algorithm is proposed to solve large traveling salesman problems to optimality. Computational experience in solving ten problems of between 48 and 318 cities is given. The algorithm combines branch and bound with constraint generation. The constraints, that are added as they are needed, are facets of the polytope associated with the problem. To test their ideas the authors combined their own code for generating facets with the IBM code MIP/370.

Addresses of Authors and Editors

N. Christofides
Department of Management Science
Imperial College of Science and Technology
Exhibition Road
London SW7 2BX, U.K.

M. Grötschel
Mathematisches Institut
Universität Augsburg
Memminger Strasse 6
D-8900 Augsburg, F.R.G.

R.M. Karp, E.L. Lawler
Computer Science Division
University of California
Berkeley, CA 94720, U.S.A.

G.A.P. Kindervater, J.K. Lenstra
Centre for Mathematics and Computer Science
P.O. Box 4079
1009 AB Amsterdam, Netherlands

F. Maffioli
Dipartimento di Elettronica
Politecnico di Milano
Piazza Leonardo da Vinci 32
20133 Milano, Italy

C.J.H. McDiarmid
Wolfson College
Oxford OX2 6UD, U.K.

G.L. Nemhauser, L.E. Trotter, Jr.
School of Operations Research and Industrial Engineering
Cornell University
Ithaca, NY 14853, U.S.A.

M. O'hEigeartaigh
National Institute for Higher Education
Glasnevin
Dublin 9, Ireland

C.H. Papadimitriou
Department of Computer Science
National Technical University
9 Heroon Polytechneiou Street
Zografou 624
Athens, Greece
Computer Science Department
Stanford University
Stanford, CA 94305, U.S.A.

S. Powell
The London School of Economics and Political Science
Houghton Street
London WC2A 2AE, U.K.

A.H.G. Rinnooy Kan
Econometric Institute
Erasmus University
P.O. Box 1738
3000 DR Rotterdam, Netherlands

M.G. Speranza, C. Vercellis
Dipartimento di Matematica
Università di Milano
Via Cicognara 7
20129 Milano, Italy

R.T. Wong
Krannert Graduate School of Management
Purdue University
West Lafayette, IN 47907, U.S.A.

Author Index